Flowering Plants of Massachusetts

BY VERNON AHMADJIAN

DRAWINGS BY BARRY MOSER

AMHERST 1979

Flowering Plants of Massachusetts

UNIVERSITY OF MASSACHUSETTS PRESS

Library of Congress Cataloging in Publication Data
Ahmadjian, Vernon.
Flowering plants of Massachusetts.
Bibliography: p.
Includes index.
1. Botany—Massachusetts.   I. Title.
QK166.A35   582'.13'09744   78-19690
ISBN 0-87023-265-7

This book is dedicated to the late David Potter, Professor at Clark University from 1924 to 1959, and to all the amateur botanists of Massachusetts. In 1931, David Potter founded the Hadwen Botanical Club which consisted of a group of dedicated amateurs who systematically collected flowering plants throughout Worcester County. The minutes of the meetings of this club reflect a zeal and enthusiasm for collecting and identifying plants that seems to be the special trademark of all amateur botanists.

Contents

Introduction

This book is intended as a guide to the common flowering plants of Massachusetts. It includes trees and shrubs as well as the more common and conspicuous wildflowers. Sedges and grasses also are included because they represent such a large segment of flowering plants; however, only a few of their species are described to illustrate these important but taxonomically difficult families.

One may question the need for such a book in a state that has been so well botanized. Since there are already several detailed and comprehensive floras which include this region, as well as a number of popular guides to the wildflowers of New England and the Northeast (see Additional References), why then another one? There are several answers to this question. First, a book that focuses solely on plants that commonly grow in Massachusetts, a text that is not merely a part of a broader coverage, can help one further narrow the options or choices when trying to identify a particular plant. Second, this book, although a popular guide, is more advanced than the extremely simple treatments of some popular flower books, yet it does not contain the extensive technical detail of terminology and species variation found in the more scientific treatises. Thus, the book bridges the gap between the overly simple and the professional manuals. It will be useful to knowledgeable amateurs who are beyond the treatments found in the popular flower guides and yet are not sufficiently experienced to handle the terminology and subtle species distinctions of the professional manuals. This book will satisfy many individuals in terms of the range of information they desire. For others it will be a starting point for more serious studies using the professional guides. It will serve both as a guide for identifying unknown plants and as a means of making the reader aware of the wide variety of common flowering plants which occur in Massachusetts.

The species included in this book are grouped within families. Except for maintaining the natural separation of monocotyledons and dicotyledons, families are listed alphabetically, as a matter of convenience. For each family of plants its more distinguishing characteristics are given. Detailed descriptions are provided for those species that are illustrated. Species common in Massachusetts but for which there is no illustration are described more briefly in relation to the illustrated species. For example, while *Lilium philadelphicum* (wood lily) is both illustrated and described, other common lilies such as *L. canadense* (wild yellow lily) and *L. superbum* (Turk's-cap lily) are also described. For the maple family, there is only one illustration—that of *Acer saccharum* (sugar maple)—and yet there are six other species of maples found commonly in the state. These six species are included under the maple family but with briefer descriptions than that of the primary, illustrated example. The same applies for asters and goldenrods: one common species is illustrated as an example of the family, but a dozen or more other common species of these plants are also briefly described. Where there are many species, as in the asters and goldenrods, identification of plants is difficult because all the species are not illustrated. Therefore, the reader must carefully consider the differences between the species and, by process of elimination, select the correct one. In some groups of plants the distinctions between species are not clear or they are based on small morphological differences. In this book, as far as possible, related species are included only if the differences are clear.

This book is not intended to be a comprehensive treatment of the flowering plants of Massachusetts. The reader certainly will find plants that are not described in this book, at which time other manuals listed in the reference section should be consulted. If a plant cannot be found in this book, it was left out either because it was not considered common or because the family to which the plant belongs was not included. There are 277 plants illustrated in these pages but the total number described is 495. These plants are representatives of 93 families.

Designating what is common is sometimes difficult and necessarily arbitrary. In general, such distinctions were made following Seymour (1969). In some families, plants which are not common have been included because they are such an interesting member of that family (i.e., the golden club in the arum family).

Many families of flowering plants have a wide distribution throughout the earth, with most species growing in one particular region. For example, while the arum and ginseng families have a primarily tropical distribution, they have common representatives in our state. Descriptions of such families are based mainly on the species that grow in Massachusetts rather than on the diverse range of species within these families.

Most of the species descriptions in this book were based on specimens observed in their natural habitats or from recently collected herbarium specimens. I wish to acknowledge especially the recent collections made by Jacqueline John and Sue Gallagher and those made by Cyrus Darling. Extensive use has also been made of the collections of the Clark University Herbarium. These collections were made over the past thirty years by the former Hadwen Botanical Club and by Dr. and Mrs. Burton Gates from the different townships of Worcester County.

The designations of height are the maximum reached by a particular plant. This may be confusing since many plants in a particular region do not reach the maximum height recorded for other individuals of this species; this value, however, gives an indication of the relative height differences among plants.

For trees and shrubs, the figures for height, in general, follow those given by Petrides (1958). Another aspect of confusion is the flowering time. Those listed are generally the periods of peak bloom, but many plants in isolated instances may mature earlier or later than most members of their species.

The information on habitat was selected from personal experience in the field, from Seymour (1969), and from the collecting data on herbarium sheets in the Clark University Herbarium.

Many species of flowering plants are found in swamps and bogs, and around lakes and ponds. There are many of these interesting habitats throughout Massachusetts; information about their locations can be obtained from McCann et al. 1972.

Since plants do not recognize state boundaries, many of the plants described in this guide have a distribution outside of the state.

Those familiar with identifying wild flowers will understand that the application of common names to plants does not follow a uniform system. Thus, while one plant may have many different common names, the one used in any work depends on the preferences of the author. So it is with this book. The common names were selected with regard to their appropriateness for a particular plant and because of regional familiarity. The Latin names, for the most part, follow those in *The New Britton and Brown Illustrated Flora of the Northeastern United States and Adjacent Canada* (Gleason, 1968).

I wish to gratefully acknowledge the assistance of the following individuals: Harry Ahles, who supervised the drawings of the different plants and gave valuable advice on the families to include in the book; Janice M. Sweeney, for her companionship on numerous field excursions and for patient proofreading; Katherine C. Hildreth, for her help in designing the keys; Margaret B. Ives, for her curatorial assistance in the herbarium; Elizabeth M. Rogers and Rene Baril, for their skillful typing; Alice C. Higgins and Rudoph F. Nunnemacher for their efforts to secure funding to publish this book; Caroline E. Woolner for advice on the glossary illustrations; Margaret Mirick, for sharing her extensive knowledge of the mint family; John W. Copithorne, Superintendent of Quabbin Section, for his permission to collect plants within the Quabbin Reservoir Reservation. The illustrations for this book were financed by the David Potter Fund, established by former students and friends of David Potter.

Keys to the Families of Flowering Plants Included in This Book

An unknown plant may be identified by several methods. The easiest way is to look through an illustrated guide until one finds a match between the specimen and a picture in the book. Certainly, if one has a general idea of the type of plant collected, this is an efficient and convenient method. Another way is to search through a herbarium—a collection of dried and pressed plants—until one finds a suitable match. Again, it helps to have a general familiarity with the plant. Since most people do not have ready access to a herbarium collection, this method is impractical. A third method of identification is to use a key, a device by which the user makes choices concerning the basic characteristics of a plant. Each choice leads to another pair of choices and so on until, hopefully, a final choice is reached and the plant is correctly identified. (For example, using the key in this book, if the plant in question is determined to be a tree, the next step is to find number 2 in the key and make a further determination. If one decides, however, that shrubs and woody vines is the correct classification, the next number to consult would be 16.)

Most keys are artificial because they do not reflect the natural relationships of plants. Keys are devices of convenience that give the user a reasonable chance to identify an unknown plant or at least come close enough to establish its morphological relationships. The keys in most professional books are so detailed that only the most skilled observer can manipulate them. In this book, the key includes only those families of plants described in the book, in the belief that such a broad key will make it easier for the reader to choose the particular family to which an unknown plant belongs. Once the family has been identified, the number of choices within each family is not so vast as to prevent the reader from identifying the plant.

One way to distinguish between flowering plants—a separation upon which this key is based—is to distinguish between monocotyledons and dicotyledons. Plants that belong to the monocotyledons have flower parts in sets of three or multiples of three, and the leaves of many species have parallel veins. (There are exceptions. For example, the leaves of *Trillium* and jack-in-the-pulpit have a network of veins.) Plants belonging to the dicotyledons have flower parts in sets of four or five or their multiples, and their leaves have a network of veins. (There are exceptions here as well: *Plantago* has parallel-veined leaves.)

Key

A. Trees, Shrubs, and Woody Vines
B. Herbs
   I. Monocotyledons
   II. Dicotyledons

A. TREES, SHRUBS, AND WOODY VINES

1a. Trees—2
1b. Shrubs and woody vines—16
2a. Leaves compound—3
2b. Leaves simple—6
3a. Flowers in catkins; fruit a nut—*Juglandaceae* (p. 311)
3b. Flowers not in catkins—4
4a. Fruit a twisted samara—*Simaroubaceae* (Tree-of-Heaven) (p. 513)
4b. Fruit a samara, but not twisted—5
5a. Fruit one long, straight samara—*Oleaceae* (White ash) (p. 378)
5b. Fruit a double samara—*Aceraceae* (Ash-leaved maple) (p. 91)
6a. Leaves opposite—7
6b. Leaves alternate—8
7a. Leaves with prominent lobes—*Aceraceae* (p. 91)
7b. Leaves without lobes—*Cornaceae* (p. 220)
8a. Leaves with unequal bases—9
8b. Leaves with equal bases—11

9a. Flower stalk attached to a leafy bract—*Tiliaceae* (Linden) (p. 525)

9b. Flower stalk not attached to a bract—10

10a. Flowers appear in spring; without petals; fruit is a samara —*Ulmaceae* (p. 527)

10b. Flowers appear in fall; petals threadlike; fruit a woody capsule—*Hamamelidaceae* (Witch hazel) (p. 305)

11a. Bark greenish-white and mottled—*Platanaceae* (Sycamore) (403)

11b. Bark not greenish-white and not mottled—12

12a. Sap milky—*Moraceae* (Mulberry) (p. 369)

12b. Sap not milky—13

13a. Flowers arranged in catkins—14

13b. Flowers not arranged in catkins—*Cornaceae* (Alternate-leaved dogwood) (p.220)

14a. Seeds and catkins with long, soft hairs—*Salicaceae* (p. 469)

14b. Seeds and catkins without hairs—15

15a. Fruit (nut) within or only partially enclosed by a coarse husk—*Fagaceae* (p. 286)

15b. Fruit (nut) not within a husk—*Betulaceae* (p. 125)

16a. Vines—17

16b. Shrubs—19

17a. Climbing by means of tendrils—*Vitaceae* (p. 557)

17b. Vines without tendrils—18

18a. Flowers small and green; fruit an orange capsule with red seeds—*Celastraceae* (Bittersweet) (p. 155)

18b. Flowers not small; light purple; fruit a berry—*Solanaceae* (p. 515)

19a. Leaves opposite or whorled—20

19b. Leaves alternate—22

20a. Flowers on ball-shaped heads—*Rubiaceae* (Buttonbush) (p. 461)

20b. Flowers not on ball-shaped heads—21

21a. Flowers axillary—*Lythraceae* (p. 357)

B. HERBS

I. Monocotyledons

1a. Plants grasslike or reeds, with dry, inconspicuous flowers
—2
1b. Plants not grasslike, most with noticeable flowers—6
2a. Stems triangular—*Cyperaceae* (p. 21)
2b. Stems round—3
3a. Flowers in compact, spongy masses—*Typhaceae*
(Cat-o-nine tails) (p. 85)
3b. Flowers not in spongy masses—4
4a. Flowers packed into round heads—*Sparganiaceae* (p. 82)
4b. Flowers not packed into round heads—5
5a. Leaves forming an open sheath around the stem; fruit a
seedlike achene—*Gramineae* (p. 25)
5b. Leaves not forming an open sheath around the stem; fruit
a capsule—*Juncaceae* (p. 31)
6a. Inflorescence a spadix with enveloping spathe—*Araceae*
(p. 8)
6b. Inflorescence not a spadix—7
7a. Plants growing in shallow water—8
7b. Plants mostly terrestrial, some found in bogs—9
8a. Flowers with white petals—*Alismaceae* (p. 3)
8b. Flowers with blue petals—*Pontederiaceae* (p. 80)
9a. Flowers irregular—10
9b. Flowers regular—11
10a. Flowers with one or two stamens fused with the style and
one petal that is highly modified to form a lower lip—
Orchidaceae (p. 57)
10b. Flowers with three stamens and no lower lip—*Iridaceae*
(p. 26)
11a. Flowers associated with prominent bracts—12
11b. Flowers not associated with prominent bracts—13

II. Dicotyledons

11a. Inflorescence an umbel–*Asclepiadaceae* (p. 114)

11b. Inflorescence not an umbel–12

12a. Flowers unisexual and reduced, with no petals or sepals; colored bracts–*Euphorbiaceae* (p. 275)

12b. Flowers bisexual and not reduced–13

13a. Flowers irregular, with an upper and lower lip– *Lobeliaceae* (p. 355)

13b. Flowers regular, without lips–*Apocynaceae* (p. 103)

14a. Plants without chlorophyll–15

14b. Plants with chlorophyll–17

15a. Stems twining–*Convolvulaceae* (Dodder) (p. 218)

15b. Stems not twining–16

16a. Flower with a two-lipped corolla–*Orobanchaceae* (p. 387)

16b. Corolla not two-lipped–*Ericaceae* (Indian pipe) (p. 268)

17a. Irregular flowers–18

17b. Regular flowers–26

18a. Stems square–19

18b. Stems not square–21

19a. Flowers with four fused petals–*Rubiaceae* (p. 461)

19b. Flowers with five fused petals–20

20a. Plants aromatic with flowers generally in small, axillary clusters–*Labiatae* (p. 313)

20b. Plants not aromatic and flowers in terminal spikes– *Verbenaceae* (Blue vervain) (p. 550)

21a. Flowers dangling at the ends of stalks–*Balsaminaceae* (p. 118)

21b. Flowers not dangling at the ends of stalks–22

22a. Leaves compound–23

22b. Leaves not compound–24

23a. Outer petals form inflated pouches; fruit is a capsule– *Fumariaceae* (p. 293)

23b. Outer petals are wing-shaped and do not form pouches; fruit is a pod–*Fabaceae* (p. 279)

24a. Flower with an upper and a lower lip–*Scrophulariaceae* (p. 479)

24b. Flower without lips—25
25a. Lower petal extends backward and forms a pouch—
Violaceae (p. 553)
25b. Lower petal does not form a pouch—*Polygalaceae* (p. 405)
26a. Inflorescence an umbel—27
26b. Inflorescence not an umbel—28
27a. Leaf petiole forms a sheath around the stem—*Umbelliferae*
(p. 528)
27b. Leaf petiole does not form a sheath around the stem—
Araliaceae (p. 108)
28a. Nodes of stem swollen—29
28b. Nodes of stem not swollen—30
29a. Leaves opposite, without stipules—*Caryophyllaceae*
(p. 146)
29b. Leaves alternate, with stipules—*Polygonaceae* (p. 407)
30a. Petals fused to some degree—31
30b. Petals not fused—40
31a. Flowers at the base of the plant—*Aristolochiaceae* (p. 113)
31b. Flowers not at the base of the plant—32
32a. Flowers borne only along one side of the stem—
Boraginaceae (p. 128)
32b. Flowers not restricted to one side of the stem—33
33a. Leaves leathery and evergreen—*Ericaceae* (p. 251)
33b. Leaves not leathery and not evergreen—34
34a. Stamens three—*Valerianaceae* (p. 549)
34b. Stamens more than three—35
35a. Stamens four, two long and two short; stems trailing—
Caprifoliaceae (Twin-flower) (p. 139)
35b. Stamens of equal length; stems not trailing—36
36a. Stamens with wide hairy filaments—*Campanulaceae*
(p. 130)
36b. Stamens with hairless filaments—37
37a. Stamens opposite the petals—*Primulaceae* (p. 416)
37b. Stamens not opposite the petals—38
38a. Stipules present—*Rubiaceae* (p. 461)

38b. Stipules absent—39

39a. Leaves opposite or basal—*Gentianaceae* (p. 295)

39b. Leaves alternate—*Solanaceae* (p. 515)

40a. Stamens many—41

40b. Stamens six or fewer—53

41a. Leaves and flowers with black glands—*Hypericaceae* (p. 307)

41b. Leaves and flowers without black glands—42

42a. Leaves cloverlike, folded when young—*Oxalidaceae* (p. 393)

42b. Leaves not cloverlike, and not folded—43

43a. Stamens fused together by their filaments—*Malvaceae* (p. 361)

43b. Stamens not fused—44

44a. Leaves compound—45

44b. Leaves simple—47

45a. Stipules absent—*Ranunculaceae* (p. 425)

45b. Stipules present—46

46a. Receptacle expanded into a cup or mound-shaped hypanthium; leaves alternate—*Rosaceae* (p. 449)

46b. Receptacle not expanded; leaves opposite—*Geraniaceae* (p. 303)

47a. Leaves fleshy—*Crassulaceae* (p. 225)

47b. Leaves not fleshy—48

48a. Leaves bristly and stems ridged—*Melastomaceae* (p. 367)

48b. Leaves not bristly and stems not ridged—49

49a. Flowers axillary—*Lythraceae* (p. 357)

49b. Flowers not axillary—50

50a. Aquatic herbs; ovary large and thick—*Nymphaeaceae* (p. 377)

50b. Terrestrial herbs; ovary not large and thick—51

51a. Hypanthium long and tubular—*Onagraceae* (p. 381)

51b. Hypanthium not long and tubular—52

52a. Flowers small, greenish-white in long racemes—*Phytolaccaceae* (p. 399)

52b. Flowers yellow, borne singly or in small groups (growing in sandy, coastal areas)—*Cistaceae* (p. 159)

53a. Plants covered with mealy granules—*Chenopodiaceae* (p. 157)

53b. Plants not covered with mealy granules—54

54a. Petals arranged in form of a cross; four long stamens and two short stamens—*Cruciferae* (p. 227)

54b. Petals not arranged in form of a cross; stamens of equal length—55

55a. Leaves compound—*Berberidaceae* (Blue cohosh) (p. 123)

55b. Leaves simple—56

56a. Flowers small, encased in pointed bracts—*Amaranthaceae* (p. 95)

56b. Flowers not encased in bracts—57

57a. Petals white and pink striped—*Portulacaceae* (Spring beauty) (p. 415)

57b. Petals not white and pink striped—58

58a. Flowers tiny, green, and unisexual—*Urticaceae* (p. 539)

58b. Flowers not green—59

59a. Flowers with a cup-shaped hypanthium around the ovary —*Santalaceae* (p. 473)

59b. Flowers without a cup-shaped hypanthium—*Linaceae* (p. 353)

Monocotyledons

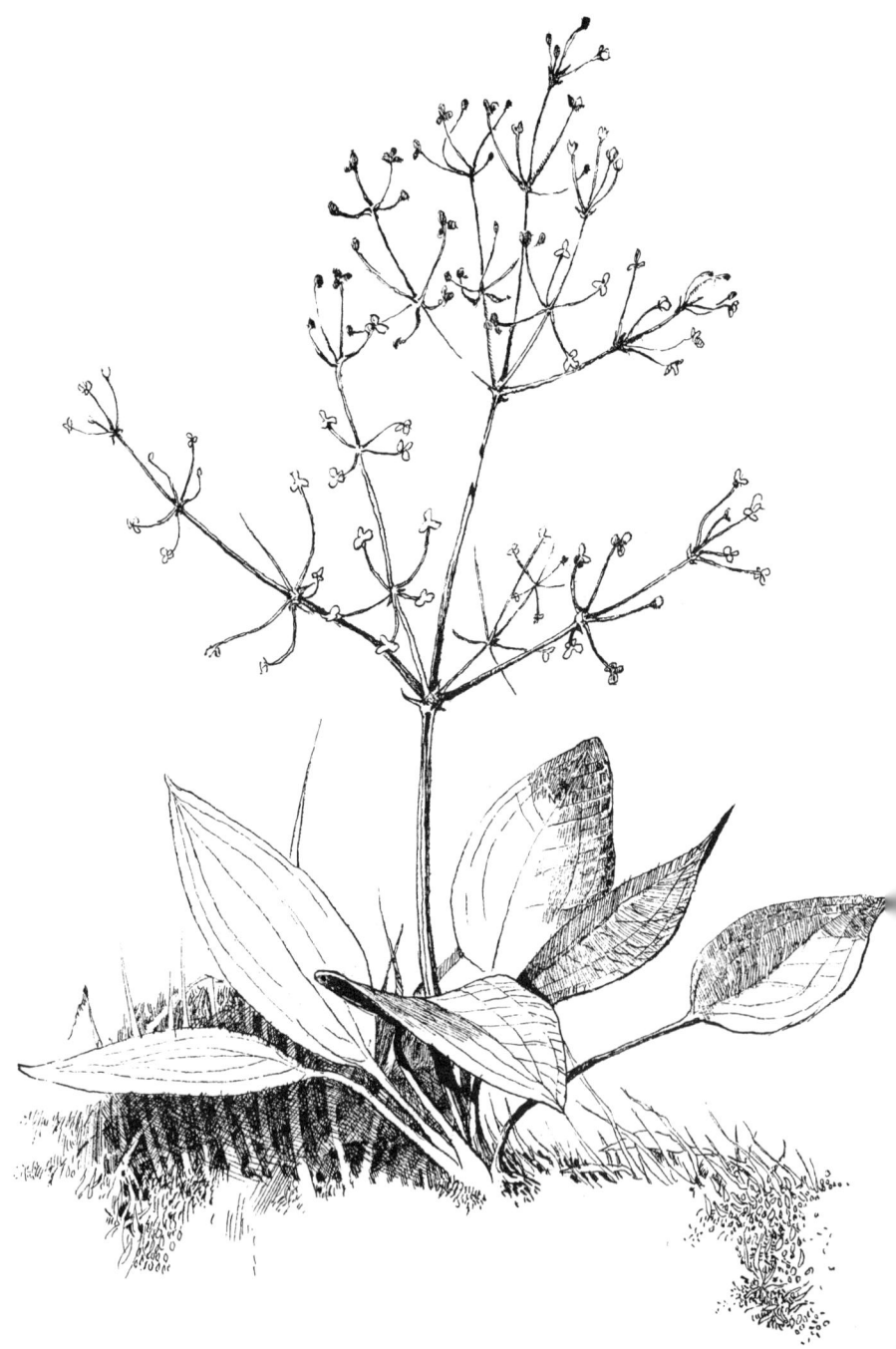

ALISMACEAE   *Water-plantain family*

This small family of plants grows in shallow water along the shores of rivers, ponds, and lakes, and in marshes and bogs. The flowers are either unisexual or bisexual, on separate stalks, and they have three green sepals and three white petals. Usually, there are many stamens and pistils and the fruit is an achene.

ALISMA TRIVIALE
Water Plantain
Wet areas / June–September / 3 ft.

From the midst of the elliptical leaves of this plant arises a tall flower stalk with small white or pink flowers at the ends of many small branches. If the plant is growing in deep water, its leaves are much narrower.

SAGITTARIA LATIFOLIA
Arrow-head
Wet areas / July–September / 3 ft.

The leaves of this plant vary in width, have a characteristic arrow shape, and protrude above the water. The white flowers are in groups of three on short stalks which grow out of the main stem of the flower stalk. The uppermost flowers are generally staminate or perfect, and the lower ones are pistillate. The pistillate flowers usually open and mature before the upper staminate flowers have opened, a feature that facilitates cross-pollination. *Peltandra virginica* (arrow arum) is another common plant of marshes and swamps with arrow-shaped leaves. It belongs to the arum family and it has a narrow spadix enveloped by a green spathe.

AMARYLLIDACEAE   *Amaryllis family*

The traits of this family are similar to those of the lily family; that is, the flower has three petals, three petaloid sepals, and six stamens. The principal difference is that the position of the ovary is inferior in the amaryllis family. An inferior ovary means that all the other parts of the flower are attached to the receptacle above the ovary. The flowers are borne on a separate stalk from the leaves, as in the daffodil, and the fruit is generally a capsule.

HYPOXIS HIRSUTA
Yellow Star Grass
Fields and open woods / May–July / 12 in.

The yellow star grass is a hairy plant with a cluster of grasslike leaves. Thin flower stalks emerge from within these leaves. Each stalk bears a small group (an umbel) of yellow, star-shaped flowers.

ARACEAE   *Arum family*

This is a family of aquatic herbs with a distinctive inflorescence. The individual flowers are small and crowded together on a fleshy stalk, called a spadix. The spadix may be round or club-shaped. It is associated with a specialized leaf, called a spathe, which is hood-shaped or petallike. Generally, the fruit is a berry, but in some species it may be dry.

ACORUS CALAMUS
Sweet Flag
Swampy areas / June–July / 5 ft.

The inflorescence (spadix) of sweet flag seems to emerge from the side of a stem which resembles its grasslike leaves, but it is separate. Actually, the inflorescence sits at the end of a stalk, and a narrow, long (a foot or more) spathe similar to the stalk projects beyond the inflorescence. The tiny flowers are clustered on a long, cylindrical spadix. The flowers are perfect, with six sepals, six stamens, and a pistil. The leaves and rhizome are aromatic.

ARISAEMA ATRORUBENS
Jack-in-the-Pulpit
Wet woods / May / 3 ft.

The spathe of this easily recognized plant has a distinctive
hood shape and it envelops a long clublike spadix. The lower
part or handle of the spadix contains the small flowers. The
spathe has alternating green and purple-brown stripes, but the
degree and extent of the different colors varies. Some plants
have mostly green spathes while others have purple-brown
spathes. The flowers are unisexual; i.e., either staminate or
pistillate. Some plants have both types of flowers, the pistillate clustered around the bottom of the spadix and the staminate flowers above them. Other plants have only staminate or
pistillate flowers on the spadix. Some predominantly pistillate-
flowered or staminate-flowered forms may have a few flowers
of the opposite type. The green ovaries develop into bright
scarlet berries which are revealed when the spathe withers.
The underground stem is tuberlike, which is why the plant is
often called Indian turnip. The tuber should not be eaten because it contains high concentrations of irritating crystals
which are difficult to remove even after boiling. There are
usually two leaves, each subdivided into three leaflets.

CALLA PALUSTRIS
Water Arum
Swampy areas / June–July / 12 in.

The distinctive white spathe of the water arum is large and petallike, and it forms a background for the short spadix. Small flowers cover the entire surface of the spadix, unlike jack-in-the-pulpit where the flowers occur only on the basal part of its long spadix. The fruit is a berry. The leaves and flower stalk grow out from a long rhizome which also gives rise to fibrous roots.

ORONTIUM AQUATICUM
Golden Club
Rare, ponds along the coast, inland bogs / May–June / 12 in.

This plant is an unusual member of the arum family. A small
spathe is located at the base of the flower stalk and it falls off
soon after the flowers appear. The small, bisexual flowers are
crowded on a club-shaped spadix. Each flower has four to six
yellow sepals, four to six stamens, and one stubby pistil partly
embedded in the fleshy tissue of the spadix. The leaves are
on long stalks and they may float on the water if it is deep
enough.

SYMPLOCARPUS FOETIDUS
Skunk Cabbage
Along woodland brooks, in swamps and wet meadows
March–April / 2 ft.

The inflorescence of the skunk cabbage appears in early spring, often melting its way through the persistent ice because of its surprisingly high temperature. The flowers appear before the leaves and they are small and clustered on a round spadix. The spadix is surrounded by a hoodlike spathe whose color is either a mixture of dark red to brownish and yellow, or green stripes and spots. Each flower is bisexual, with four fleshy sepals, one stamen opposite each sepal, and a stubby pistil whose ovary is buried in the fleshy mass of the spadix. In each flower, the tips of the stamens and pistil protrude above the closely pressed sepals. After the flowers have developed, the leaves begin to appear. They emerge from the ground as tightly rolled spears next to the flower, and later they become very broad. The crushed parts of the plant emit a strong, disagreeable odor which is the basis for the common and Latin names of this plant.

COMMELINACEAE *Spiderwort family*

The spiderwort family includes herbs with flowers that bloom only for a day. The flowers emerge from within a folded leaf-like bract and each flower has three sepals, three petals, and six stamens. The base of the leaves form a sheath around the stem. The fruit is a capsule.

COMMELINA COMMUNIS
Dayflower
Gardens, near roadsides, open lots / July–September / 12 in.

The dayflower is a creeping, weedy-type plant that roots easily from nodes which touch the ground. Each flower has two erect blue petals and a third, much smaller, lower white petal. Three of the stamens are small and sterile.

19

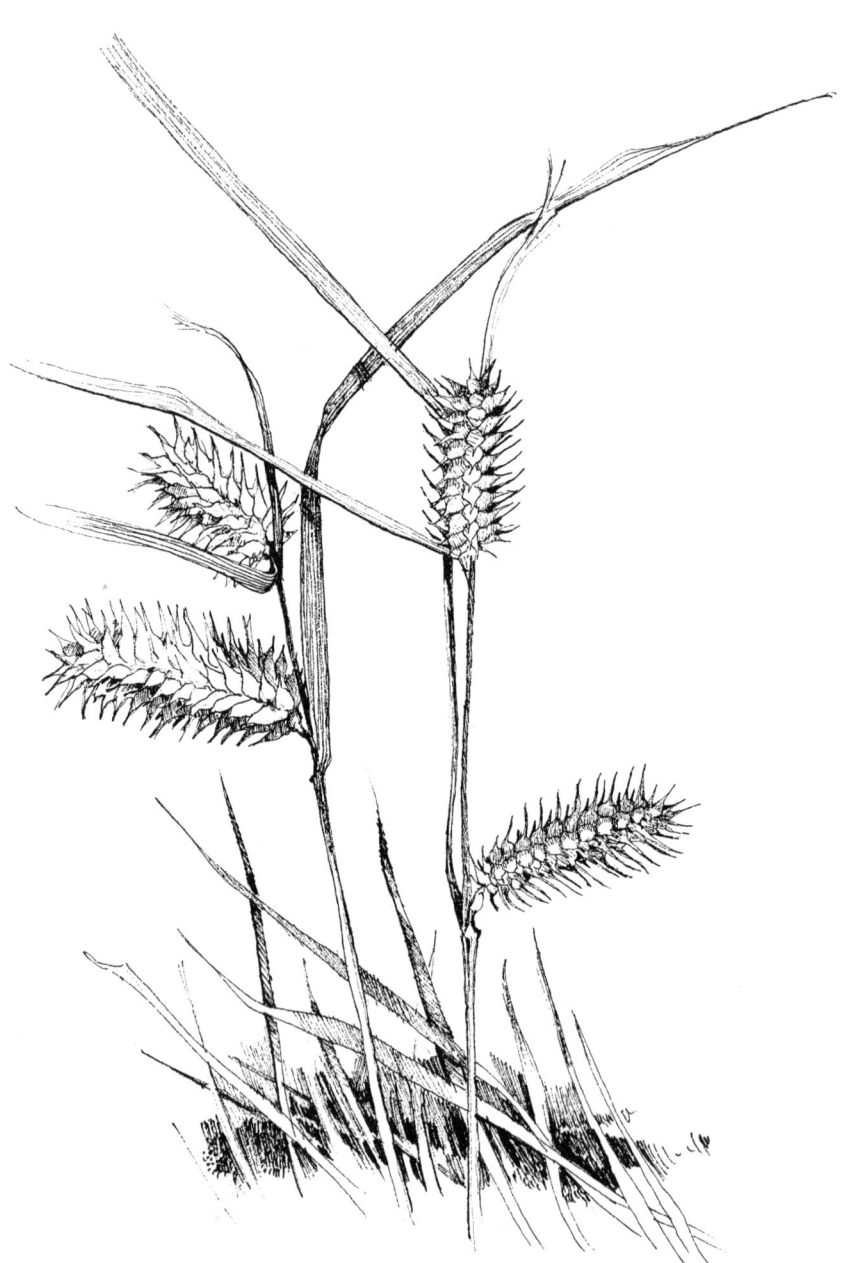

CYPERACEAE *Sedge family*

The sedges are a large family of grasslike plants. The identification of sedges and grasses is difficult even for the professional botanist. One way to distinguish between a sedge and a grass is to roll the stem of the unidentified plant between thumb and forefinger. If the stem feels triangular, it is most likely a sedge. If the stem feels round, it may be a grass. There are other points of distinction. The sedge stem is solid and its leaves form a closed sheath around the stem. In grasses the stem is hollow and the leaves form an open sheath around the stem. The flowers in both families are tiny, with sepals and petals either missing or reduced to inconspicuous bristles, hairs, or scales. The fruit is an achene. Sedges are found commonly in swamps and marshes and they flower between July and September.

CAREX LURIDA
Wet, open fields, woods / 3 ft.

Carex is the genus with the largest number of species in Massachusetts. The unisexual flowers are grouped together on spikes, either on separate staminate and pistillate spikes or on different regions of the same spike. The ovary is enclosed within a membranous sac called a perigynium, and the styles of the pistil emerge from an opening at the tip of this sac. There is one pistil with two to three styles, and each staminate flower consists of three stamens. *C. lurida* has one long staminate spike situated above two to four pistillate spikes. The perigynium has a long beak and there are numerous perigynia on the spike.

CYPERUS ESCULENTUS
Umbrella Sedge
Shores of rivers and ponds / 2 ft.

This is one of the most common of the approximately fifteen species of *Cyperus* in Massachusetts. Members of this genus are weedy plants. The inflorescence is branched and clustered together at the end of the stem and there are leafy bracts at the base of the inflorescence. The papyrus plant (*C. papyrus*) is commonly grown as an ornamental in greenhouses; its soft inner tissue was once used by the ancient Egyptians to make paper.

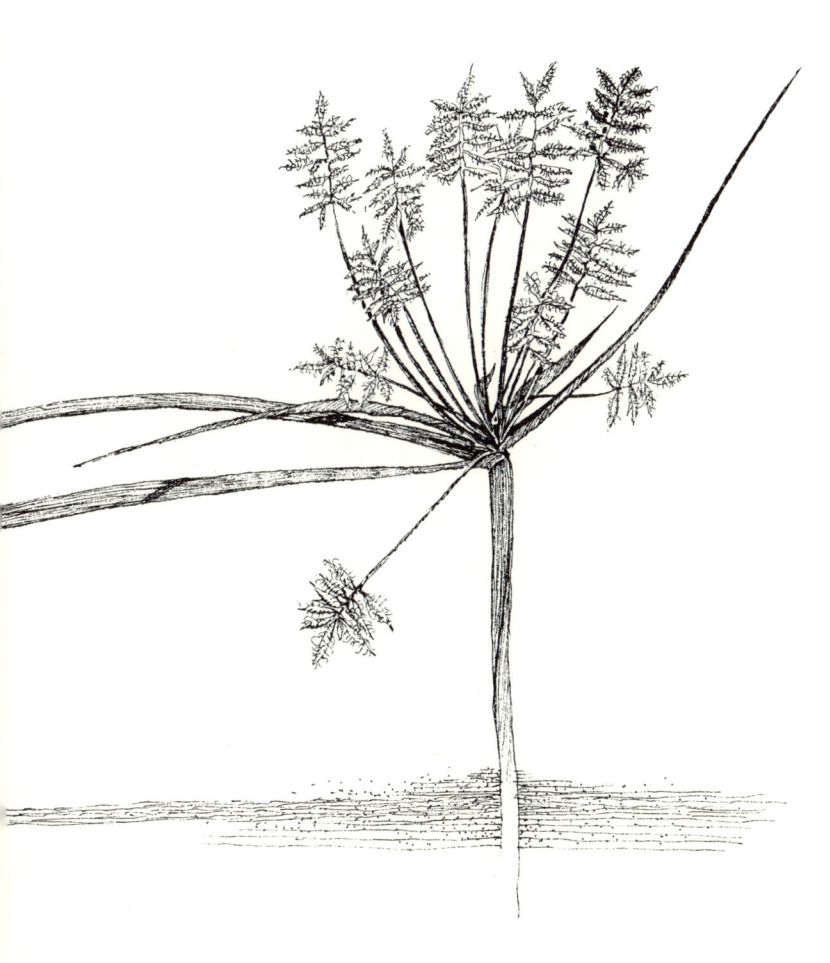

GRAMINEAE   *Grass family*

This family includes many different species of economically important plants. The flowers, called florets, are highly reduced (for most grasses, details of the florets can be seen only with a magnifying lens), and they are grouped in a basic pattern called the spikelet. Spikelets are arranged in a particular fashion to form a specific type of inflorescence, such as a spike, raceme, or panicle. A spikelet consists of one of several florets; at the base of each spikelet is a pair of bracts called glumes. A floret may be staminate, pistillate, or bisexual (having both stamens and pistils). Sepals and petals are absent, but there are specialized bracts and scales associated with the florets that may represent remnants of sepals and petals. There are two bracts on either side of a floret: the larger bract is called the lemma and in some grasses it is extended into a long bristle called an awn; the smaller bract is called the palea and is usually partially enclosed by the lemma. Because grasses are wind-pollinated, the style and stigma tend to be large and hairy in order to facilitate capture of wind-blown pollen, and the anthers tend to be long and they dangle out like threads from the mature florets. The fruit or grain is one-seeded and indehiscent. For other details of the grass family refer to the sedge family (Cyperaceae).

HYSTRIX PATULA
Bottle-brush Grass
Woods / July–August / 3 ft.

Each spikelet of this grass has two florets, and the lemmas have long awns. The spikelets are arranged into an erect spike.

25

IRIDACEAE *Iris family*

This family of herbs has flowers with three sepals, three petals, three stamens, an inferior ovary, and a style with three branches. The leaves are long, narrow, and parallel veined and the fruit is a capsule. The family includes ornamental species such as crocuses and gladioli.

IRIS VERSICOLOR
Blue Flag
Wet fields, ponds, marshes / June–July / 3 ft.

The flower of blue flag is similar to that of the cultivated irises with three petallike sepals which bend downward. The inner part of each sepal is yellowish-green with purple veins and a white border; this pattern of colors probably serves as a nectar guide for pollinating insects. The petals are narrower than the sepals and they stand upright. The three styles are petallike; they bend over the sepals, and the end of each style flares up into two lobes. Together, a sepal and a style form a type of chamber into which the insect crawls in search of nectar. The stigma is a narrow band of tissue at the end of the style where it flares up to form the flaplike lobes. Each of the three stamens is enclosed beneath the arching style. The ovary is long and narrow and has three chambers. *Iris prismatica* (slender blue flag) has very narrow leaves (less than 1 cm. wide) and a sharply three-angled ovary and seed capsule. The leaves of *I. versicolor* are wider (up to 3 cm.) and its capsule is less sharply three-angled.

SISYRINCHIUM ANGUSTIFOLIUM
Blue-eyed Grass
Fields, meadows / May–June / 18 in.

This grasslike plant has bright blue flowers with yellow centers. Each sepal and petal has a spiny tip, and the three stamens are joined to the style to form a central column. The usually branched flower stalk is separate from the narrow leaves; a pair of thin wings occurs along both sides of the stalk. *Sisyrinchium montanum* is similar, but its flower stalks are unbranched. It is a more common species. Also, the inner and outer flower bracts of *S. angustifolium* are about the same length while in *S. montanum* the outer bract is much longer than the inner bract. *S. atlanticum* grows more commonly in fields in central and eastern Massachusetts. It has purple flowers, branched stems, and very slender pale green leaves and flower stalks.

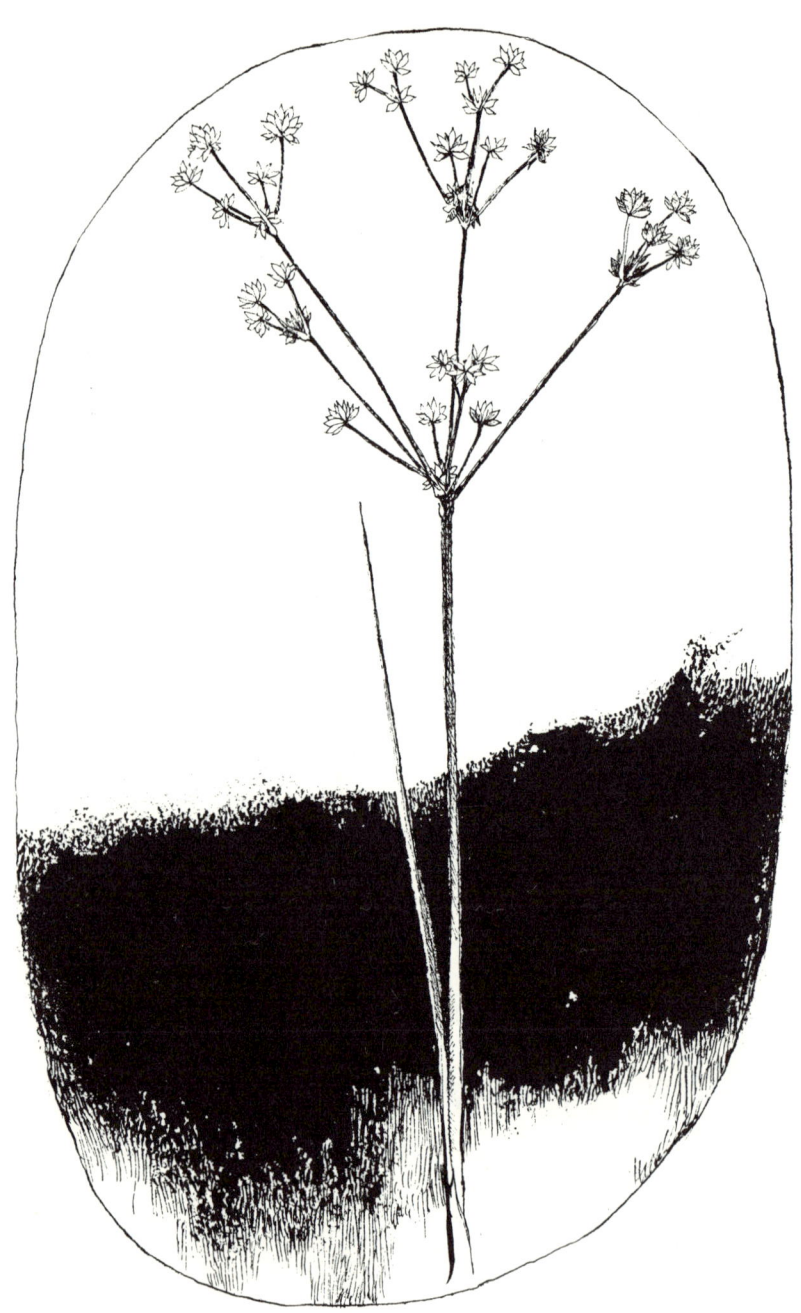

JUNCACEAE   *Rush family*

The plants in this family have reduced flowers like those of the grasses and sedges. The floral pattern for many of the rushes, however, is similar to that of the lily family: three scaly sepals, three small petals, six stamens, and one pistil. Some rushes have only three stamens. The fruit is a capsule.

JUNCUS CANADENSIS
Marsh Rush
Along streams and ponds, swamps, fields /
July–September / 3 ft.

The inflorescence of the marsh rush is terminal with erect branches bearing small groups of flowers. The leaves are hollow and smooth, and the stems grow together in stiff tufts.

JUNCUS EFFUSUS
Tufted Rush
Swamps, wet fields / July–September / 3 ft.

The inflorescence of this rush emerges from the side of a leaf-like structure. The flower-bearing branches are either widespread or compact.

LILIACEAE   *The Lily family*

This is an easy family to recognize because the flowers have a similar pattern; each flower has three sepals, three petals, six stamens, and one pistil, and the sepals and petals are the same color. There are some exceptions to this general pattern. For example, *Maianthemum canadense* (wild lily of the valley) has two sepals, two petals, and four stamens, and species of *Trillium* have sepals and petals of different colors. Representatives of the lily family are an important part of our spring flora, and lily, tulip, hyacinth, and onion relatives are well-known plants to gardeners. The species in Massachusetts are perennials that develop from underground bulbs, rhizomes, or stolons. The fruit is either a capsule or a berry with one to many seeds.

LILIUM PHILADELPHICUM
Wood Lily
Dry, open woods and clearings / Late June–August / 3 ft.

The wood lily has erect flowers that are borne either singly
or in groups of two to five. The petals and sepals are separate
from each other and their margins inroll to form a groove near
the base of the flower. Thus, each sepal and petal has a funnel-
shaped basal part and a broad outer part. The flowers are orange-red, rarely yellow, and they have roundish purple spots.
The leaves generally are in whorls of four or more at intervals
along the stem. *Lilium canadense* (wild yellow lily) has hanging flowers that are borne singly or, more commonly, in
groups of up to twenty flowers on one plant. The flowers are
yellow or orange, at times red, with reddish-brown spots. The
wild yellow lily may grow up to five feet and is found in moist,
open meadows. *Lilium superbum* (Turk's-cap lily) has flowers
with strongly recurved sepals and petals which fully expose the
stamens and pistils. There may be one or many flowers on a
stem. The flowers are orange or red with purple spots and
bright green bases. One of the largest of the lilies, Turk's-cap
sometimes grows up to ten feet. It is found in wet meadows
and woods. The fruit of a lily is a capsule.

CLINTONIA BOREALIS
Yellow Bead Lily
Woods / May–June / 15 in.

The yellow bead lily has a single flower stalk that emerges from a cluster of from two to five basal leaves. It bears an umbel of several (up to six) greenish-yellow flowers and, later in the season, dark blue berries. Large populations of this plant may develop from underground rhizomes. It grows in wet woods, commonly with the moccasin flower and wild lily of the valley.

HEMEROCALLIS FULVA
Orange Day Lily
Roadsides / May–July / 5 ft.

The day lily is commonly found—often in dense clusters—in gardens, around homes, along roadsides, and in fields where it has escaped from cultivation. The large, orange flowers have petals with crinkled margins. Each flower blooms for only one day, but new flowers are produced continually along the stem. *H. flava* (lemon day lily) is similar but has yellow, fragrant flowers. The fruit of both species is a capsule.

MEDEOLA VIRGINIANA
Indian Cucumber Root
Woods / June–July / 2 ft.

This slim, woodland plant grows from a thick, white, cucumberlike rhizome. The stem has two whorls of leaves. One whorl has five to nine leaves above the middle of the stem, and a second whorl has three to four leaves at the top of the stem. The small, greenish-yellow flowers emerge from the top whorl of leaves and hang down below the leaves. The sepals and petals curve backward and expose prominent reddish stamens. The three brown stigmas are long, widespread, and curve backward. The fruit is a dark, purplish berry.

POLYGONATUM PUBESCENS
Solomon's-seal
Woods / May–June / 3 ft.

Solomon's-seal has a gently arching stem. Its leaves are hairy on their undersurface, and its yellow-green flowers are tubular with fused sepals and petals. Flowers emerge from leaf axils and they hang downward, either singly or in pairs. The underground rhizome is thick with raised, bowl-shaped scars resembling the ancient seals of King Solomon, where the stems of previous years were attached. The fruit is a blue berry.

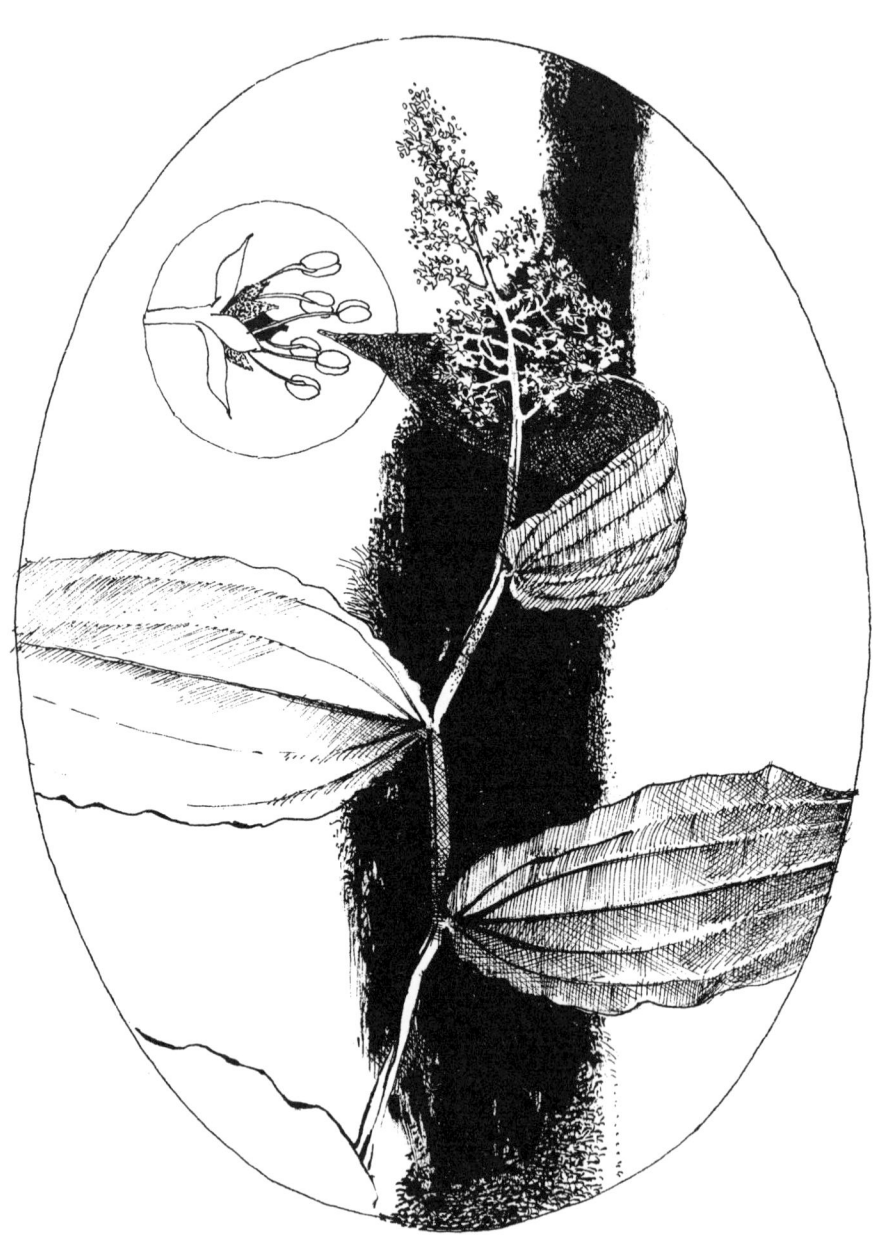

SMILACINA RACEMOSA
False Solomon's-seal
Roadsides, Woods / May–July / 3 ft.

While the common name of this plant implies a close resemblance to the true Solomon's-seal, in fact the two plants can be easily distinguished from each other. The only close resemblance between the two is in the shape of the leaves—which have prominent parallel veins—and the arching and zigzag manner of the stem. Tiny white to cream-colored flowers with very reduced sepals and petals are clustered on short branches at the end of the stem. The stamens are prominent with yellowish anthers. The fruits are small, often brown-speckled berries.

SMILAX HERBACEA
Carrion-flower
Woods / June–July / 6 ft.

This herbaceous vine has broad leaves, tendrils, and small, greenish flowers. The flowers have a disagreeable odor and are arranged in umbels at the ends of long stalks. The stems do not have thorns. The fruits are black berries. *Smilax glauca* (catbrier) has woody, thorny stems and its leaves have a white undersurface. *Smilax rotundifolia* (greenbrier) also has woody, thorny stems and broadly rounded leaves which are green on both sides.

STREPTOPUS ROSEUS
Twisted-stalk, Pink Mandarin
Woods / May–June / 2 ft.

This plant resembles Solomon's-seal in the shape of its leaves and arrangement of flowers along the stem. The stems have a zigzag pattern of growth and they usually have several branches. The pink flowers have bent stalks and they emerge opposite the leaves and then twist around toward the leaves. The fruits are red berries. *Streptopus amplexifolius* (white mandarin) has greenish-white flowers and its leaves clasp the stem.

TRILLIUM ERECTUM
Wake Robin
Woods / May–June / 18 in.

The flowers of the red trillium are usually maroon to purplish-brown but sometimes they are white, green, or yellowish. The flowers may be erect, above the whorl of three leaves, or they may be bent forward so that they hang below the leaves. The pistil and stamens are maroon-colored and the thick ovary has six winglike structures. *Trillium cernuum* (nodding trillium) has flowers which hang below and are almost hidden by the leaves. The petals curve backward and are usually white and sometimes tinged with purple or pink. The anthers and areas of the ovary are purple. *Trillium undulatum* (painted trillium) has erect flowers above the leaves. Its petals have a wavy margin, and they are white with a purple base that is streaked with crimson lines. The anthers are purple and the ovary is not winged. Fruits of these species of *Trillium* are bright red berries.

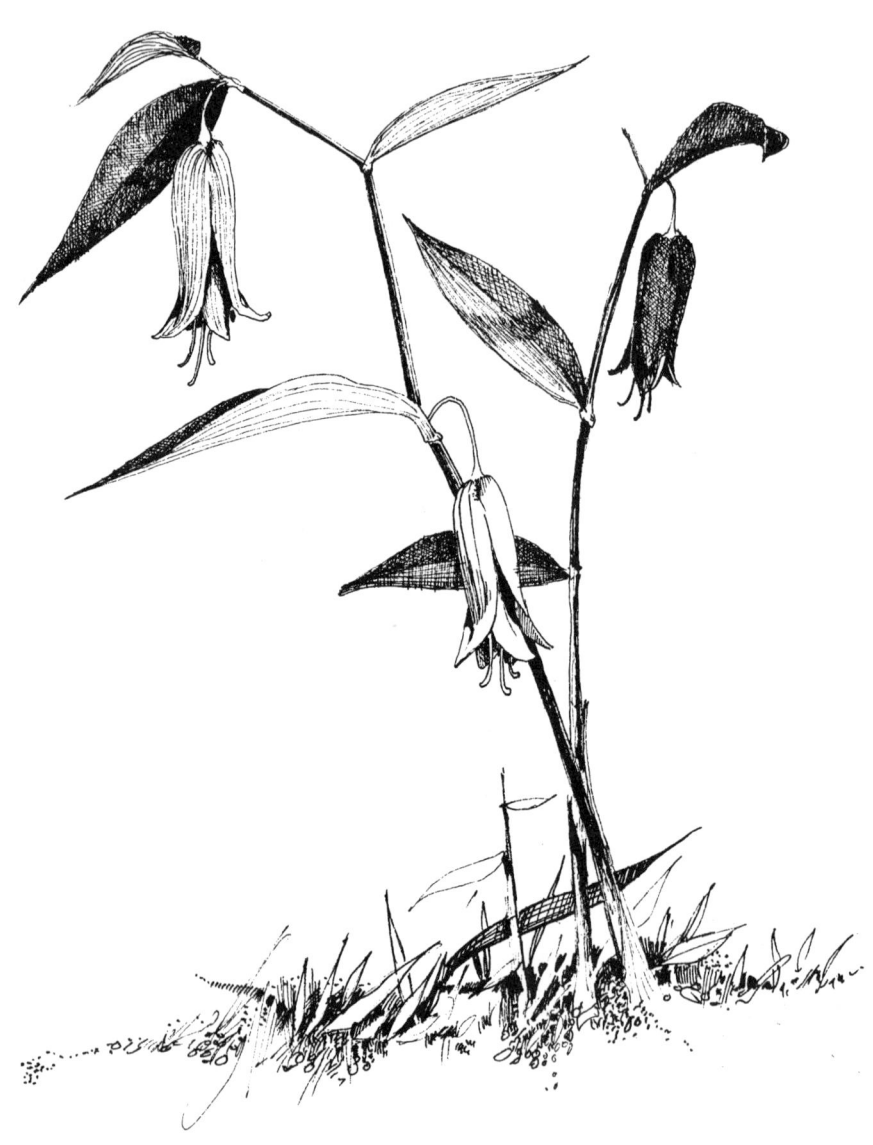

UVULARIA SESSILIFOLIA
Bellwort
Woods / April–May / 12 in.

The bellwort has upright stems which arise from a white, underground rhizome. Each stem forks at the top of the plant, bears linear leaves, and has one or two hanging yellow flowers which are narrow and bell-shaped. Anthers of the stamens are longer than the filaments. *Uvularia perfoliata* (perfoliate bellwort) has broad leaves which surround the stem in such a way that the stem appears to be growing through the leaf. The fruit of the bellwort is a capsule.

VERATRUM VIRIDE
False Hellebore
Wet fields, along streams / May–June / 6 ft.

This plant, which is also known as Indian poke, has broad, prominently veined leaves with lengthwise plaits or folds. The greenish-yellow flowers are grouped on branches at the top of the stem. The yellow anthers contrast well with the yellowish-green petals and sepals. The flower-bearing parts of the stem are hairy. The fruit is a capsule. All parts of this plant are highly poisonous.

ORCHIDACEAE *Orchid family*

The orchids are one of the largest families of flowering plants.
Although they are mostly tropical herbs, many species grow in
temperate and arctic regions. One reason for their extensive
variation in form is the orchid flower, a highly modified structure which is well adapted to cross-pollination by insects. The
flower consists of three sepals—two of them fused in some species—and three petals. Two of the petals are positioned lateral-
ly while one petal, called the lower lip, is different in shape.
The lower lip through its many modifications attracts and
serves as a landing platform for pollinators. The one or two
stamens are fused with the style to form a column which is
usually arched in the center of the flower over the lower lip.
The pollen grains stick together and form several compact
masses called pollina; these are transported intact by pollinating insects. The ovary is usually long and twisted and lies below the other parts of the flower. The fruit is a capsule with
thousands of minute seeds.

APLECTRUM HYEMALE
Adam-and-Eve
Rare, woods, mostly in western Massachusetts
May–June / 18 in.

The flowers of Adam-and-Eve are borne on a leafless stalk
which grows from an underground tuber. After the flowers
emerge, the tuber forms a second tuber from which a single
broad leaf develops in late summer. The leaf persists through
the winter and then dies before a new flower stalk emerges

from the second tuber. The close proximity of this pair of tubers is the basis for the common name of this plant. The lip of the flower is white with purplish areas while its other petals and sepals are brownish-purple.

CALOPOGON TUBEROSUS
Grass Pink
Bogs, meadows / June–July / 18 in.

The flowers of this orchid are pink and fragrant. They appear to be upside down because their lower lips stand erect at the top of the flower rather than being situated at the bottom of the flower as in most orchids. Each lip has a dense patch of yellow, pink, and orange-tipped hairs. Each flower stalk contains from three to ten flowers. A long, grasslike leaf sheaths the flower stalk.

CYPRIPEDIUM ACAULE
Moccasin Flower
Woods / May–June / 18 in.

The familiar pink lady slipper, once considered so rare that it was placed on an endangered species list, is still protected by law but it has spread abundantly throughout the woods of Massachusetts. The single flower stalk rises between a pair of basal leaves. The lower lip of the flower forms a large pouch which can be opened along its middle. The sepals and the two lateral petals are brownish-green; one sepal is erect above the pouch, while the two sepals beneath the pouch are fused and appear as one. At the top of the pouch, there are two openings, each of which contains a large, round stamen attached to a shoehornlike column which arches into the pouch. The third stamen is sterile and is a broad, leafy flap over the column. There is a green, erect, leafy bract above the flower. After the flower withers, a large, somewhat upright capsule develops at the end of the stalk.

CYPRIPEDIUM CALCEOLUS
Yellow Lady Slipper
Calcareous soil in woods / May–June / 2 ft.

This plant is similar to the pink lady slipper in many of its details. However, the pouch of the yellow lady slipper is yellow, and its lateral petals and sepals are brownish-purple or greenish-yellow. This species is found more commonly in the limestone regions of western Massachusetts. It has been reported only rarely from other parts of the state.

EPIPACTIS HELLEBORINE
Helleborine
Woods / July–September / 3 ft.

The flower stalk of this plant may be up to ten inches long and it bears numerous greenish flowers with purple veins. Each flower lies in the axil of a narrow, elongated bract. The lower lip forms an open sac and it has a tip that bends downward. The leaves are oval to lance-shaped and they lack stalks. A native of Europe, this plant has spread widely throughout the northeastern states.

GOODYERA PUBESCENS
Downy Rattlesnake Plantain
Woods / August–September / 18 in.

This evergreen plant can be recognized even without flowers because of a distinctive netlike pattern of white veins on its dark green leaves. From a circle of leaves near the ground arises a long, hairy stalk with white flowers on all sides and usually with several bracts along its length. Another species, *Goodyera repens* (creeping rattlesnake plantain), sends out runners and has flowers along only one side of the stalk. The flowers of *Goodyera tesselata* (checkered rattlesnake plantain) commonly run in a spiral along the stem; its leaves have light-green veins and a checkered pattern of light- and dark-green patches.

HABENARIA PSYCODES
Small Purple Fringed Orchid
Wet woods, along rivers / July–August / 3 ft.

The lower lip of each pink flower of the purple fringed orchid has three large lobes with deeply fringed margins. The lip curves backward into a hollow spur. There is one anther with two pollen sacs and each sac has a pollinium attached to a long stalk. The anther is fused around the stigma. The flowers are arranged in a dense raceme at the top of the stem which also bears the lance-shaped or oval leaves. The flowers of this species vary so greatly in size that a large-flowered form and a small-flowered form can be distinguished. Some botanists have recognized the large-flowered form as a separate species (*Habenaria fimbriata*; large purple fringed orchid). *Habenaria blephariglottis* (white fringed orchid) grows in peat bogs and it has racemes of crowded white flowers. Each flower has an unlobed and densely fringed lower lip with a very long spur. *Habenaria lacera* (ragged orchid) grows in meadows and has yellowish-green flowers with long spurred lower lips. The lip has three narrow and deeply fringed lobes, the middle lobe being much longer than the two lateral lobes.

ISOTRIA VERTICILLATA
Whorled Pogonia
Woods / May–June / 12 in.

The five leaves of this plant form a whorl near the top of the stem. Above this circle of leaves is a single, large flower. The flower has: three long and narrow, purplish-brown sepals; two yellowish-green, short, lateral petals; and a yellow-green lower lip with purple veins. The plants arise from long, fleshy roots which bud off new stems.

LIPARIS LILIIFOLIA
Tway-blade
Woods / June–July / 12 in.

The tway-blade has two, large, shiny leaves which arise from an underground bulb; the bulb also produces a flower stalk. The flowers are spaced along the top part of the stalk; they have narrow, greenish-white sepals and two, threadlike, lateral petals which are purple-brown. The lower lip is very broad and purplish. *Liparis loeselii* (bog tway-blade) grows in swamps and has yellow-green petals and a narrower lower lip.

MALAXIS BRACHYPODA
Adder's Mouth
Rare, swamps, western Massachusetts / June–July / 8 in.

The small, greenish-white flowers of this orchid are borne above a single, oval leaf. The lip of the flower narrows to a fine point and may have two earlike lobes at its base. This plant favors calcareous areas, but the limited reports of its collection have placed it on the list of rare plants of Massachusetts.

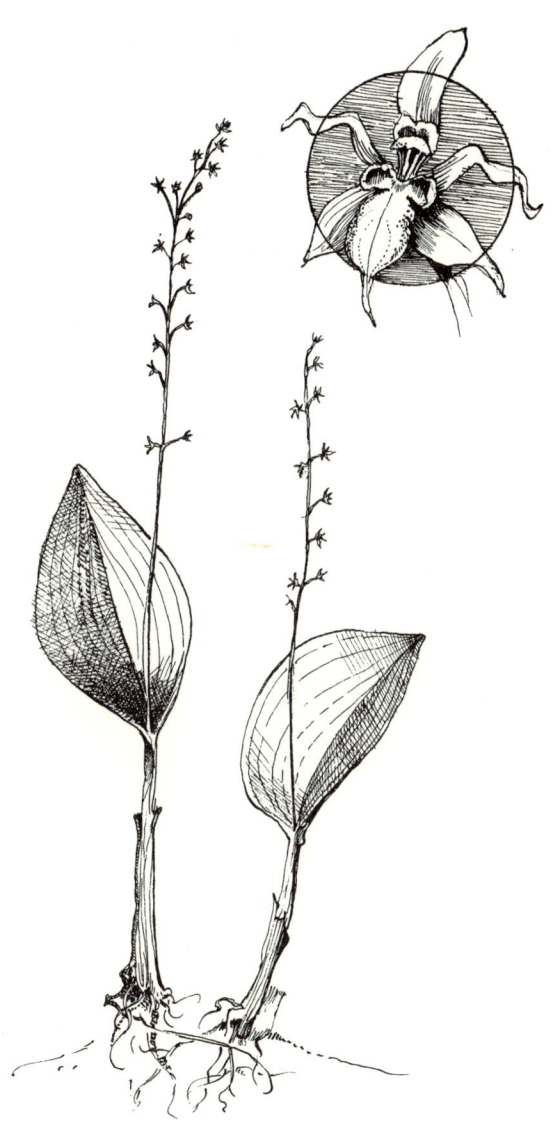

ORCHIS SPECTABILIS
Showy Orchis
Woods, calcareous soil, more common in
western Massachusetts / May–June / 12 in.

The showy orchis is a stout plant with two basal, broad leaves. The large flowers are associated with bracts. The sepals and the two lateral petals are pink to purple and stand together to form a hood which arches over the column. The lower lip is white and extends backward into a hollow spur. The flowers are arranged in a raceme on a short stem that arises between the two leaves.

SPIRANTHES CERNUA
Nodding Ladies'-tresses
Wet fields, along rivers | August-September | 2 ft.

The flowers of nodding ladies'-tresses are white, tubular, and fragrant and they are arranged in several rows along the upper part of the stem. The sepals closely embrace the petals. The much-reduced leaves occur either close against the stem or in a basal position; in the latter case, they are narrow and grass-like. The two lateral (upper) petals are fused to the upper sepal. The lip of the flower bends down at its tip and is ridged along both sides. *Spiranthes lacera* (slender ladies'-tresses) has white flowers in a single row up the stem, either on one side of the stem or in a spiral around the stem; part of the flower's lower lip is green.

PONTEDERIACEAE   *Pickerel-weed family*

This is a family of aquatic herbs, some of which have showy flowers. The water hyacinth, despite its beauty, is a troublesome weed in the navigable waters of the southern United States. The fruit has one seed and resembles an achene.

PONTEDERIA CORDATA
Pickerel-weed
Along muddy shores of ponds and rivers / July–August / 4 ft.

In Massachusetts, pickerel-weed is a common plant along the shores of rivers and ponds. Its single leaf is broad and commonly heart-shaped, and its blue flowers are grouped along the end of a stalk. Each flower consists of six, blue, petallike lobes. The lower three lobes are separate from each other while the upper three lobes are fused along half their length. The middle lobe of the upper three is the largest and has a bright yellow patch. There are six stamens fused to the tube; three of the stamens are short and they are located in the lower part of the perianth tube. There is one pistil and one long style.

SPARGANIACEAE   *Bur-reed family*

This family of aquatic herbs has grasslike leaves and unisexual flowers packed into round heads. The fruit is an achene.

SPARGANIUM AMERICANUM
Bur-reed
In water along shores of lakes, ponds,
and rivers / May–August / 3 ft.

On the individual flower stalks of bur-reed, the small staminate heads are located above the larger pistillate heads. The flowers. are very reduced. The pistillate flowers have a long style and a large, fuzzy stigma. In some plants the pistillate flowers open before the staminate flowers, a feature that promotes cross-pollination.

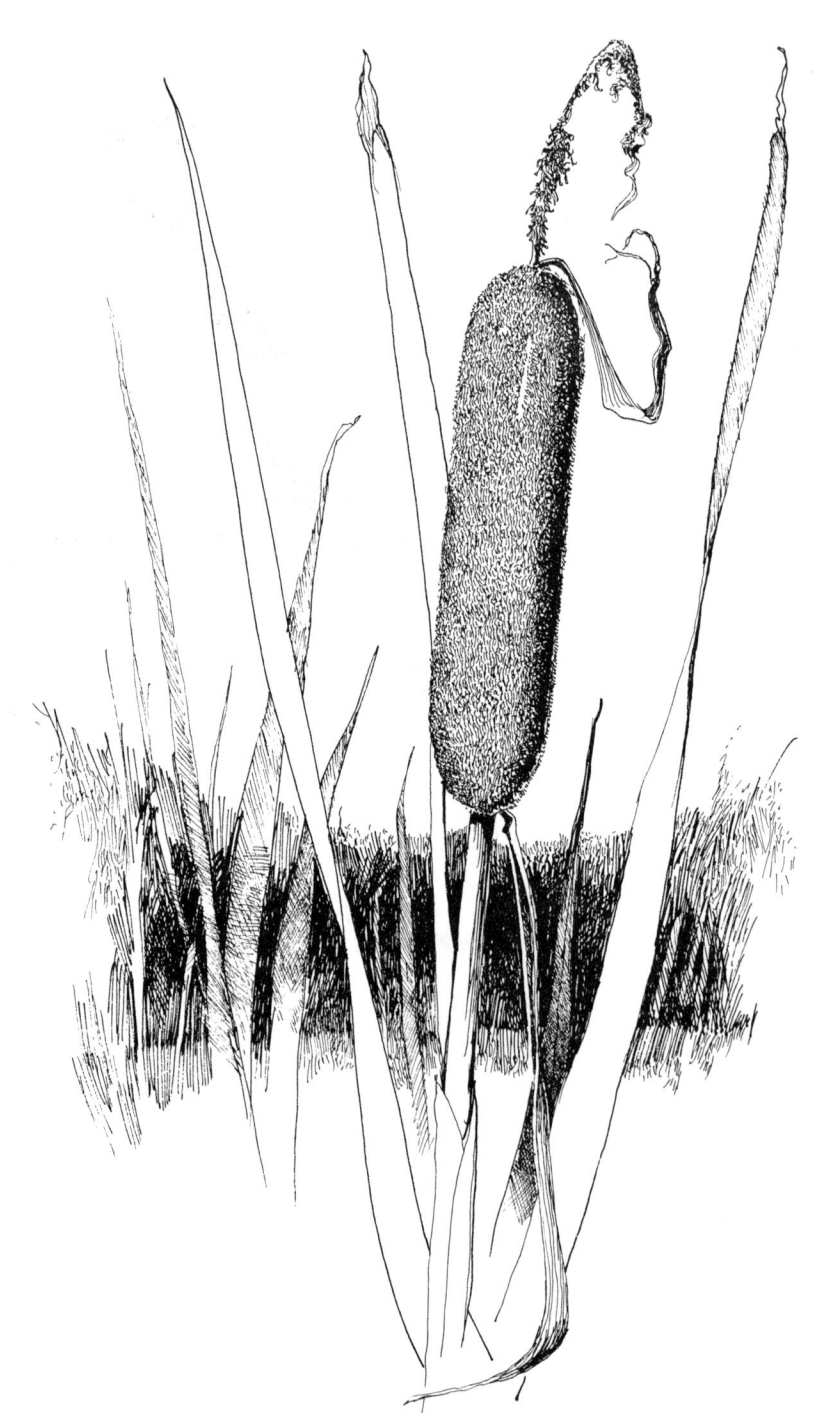

TYPHACEAE   *Cat-tail family*

Members of this small family of plants are found commonly in marshes and swamps. The cat-tail's small, reduced flowers are unisexual and are packed together in long, terminal spikes, with the staminate spike located on top of the pistillate spike. The fruit is an achene and the leaves are long and narrow.

TYPHA LATIFOLIA
Wide-leaved Cat-tail
Swamps, along ponds / July / 9 ft.

The female spike of the cat-tail is a brown, spongy, sausagelike structure that lies directly below the male spike. The male flowers are loosely clustered on their spike and fall off after shedding their pollen, leaving the central stalk to which they were attached. Both types of flowers lack sepals and petals and there is only a fringe of hair around the individual flowers. *Typha angustifolia* (narrow-leaved cat-tail) has uniformly narrow leaves and a space, about one-half inch, between the male and female spikes.

XYRIDACEAE   *Yellow-eyed Grass family*

This is a small family of mostly tropical herbs with grasslike leaves and flowers grouped in heads at the ends of separate stalks.

XYRIS CAROLINIANA
Yellow-eyed Grass
Bogs, sandy shores of ponds | July–October | 18 in.

The flower stalks of this plant are longer than the leaves. They bear yellow flowers in the axils of the tightly packed bracts which make up the head. The plant generally grows in tufts.

Dicotyledons

ACERACEAE   *Maple family*

The maple family consists of trees with opposite leaves which in most species are simple and lobed. While the ash-leaved maple (*Acer negundo*) is unusual in having compound leaves with from three to five leaflets, it does have the typical maple fruits. Flowers of these trees are small, bisexual or unisexual, and they generally have five sepals which are separate or fused. Petals may be absent; if present, they number five and are generally small and separate. Each half of the two-celled ovary develops a wing; this becomes part of the typical maple fruit, the key or samara, so that each fruit consists of two one-seeded samaras united at their bases.

ACER SACCHARUM
Sugar Maple
Woods, roadsides / April–May / 60 ft.

A commonly planted native American tree, the sugar maple
has widespread branches and a gray to black, deeply grooved
bark. Leaves have from three to five lobes with each lobe having several teeth. Flowers are small and appear in late April
and May along with the leaves. Two types of flowers—bisexual
and unisexual—occur on the same tree. Petals are absent, and
the five sepals form a tube within which the sex organs are
borne. The number of stamens varies from six to eight. *Acer
rubrum* (swamp or red maple) has a smooth, light-gray bark on
its upper branches and the typical gray, ridged bark on the older parts of the tree. Small, dark-red, bisexual flowers appear
before the leaves. *Acer saccharinum* (silver maple) is another
native American, early flowering maple whose leaves are deeply
cut into five slender lobes that are green on their upper surface
and silvery-white below. *Acer platanoides* (Norway maple) is a
commonly planted European shade tree with broad leaves and
wide-spreading samaras. Each half of the samara or key is directly opposite the other half. The other maples have samaras
that are not spread as widely (see sugar maple for illustration).
A milky juice appears at the base of the petiole when it is brok-
en off. The flowers are bisexual with eight stamens and a two-
parted style. *Acer pensylvanicum* (striped maple) is a slender,
native tree. Its distinctive green, smooth bark has pale or dark
vertical stripes and large leaves with three short lobes. *Acer
spicatum* (mountain maple) is a small, native tree with three-
lobed leaves. Its yellowish flowers appear along with the leaves
and are clustered in small groups on an erect terminal stalk.

AMARANTHACEAE   *Amaranth family*

This is a family of weedy herbs with entire, alternate leaves
and small flowers that are encased in pointed bracts. The fruit
is one-seeded and indehiscent.

AMARANTHUS HYBRIDUS
Prince's Feather
Waste areas, roadsides / July–September / 6 ft.

The flowers of prince's feather are grouped in numerous grasslike spikes which emerge from the axils of leaves along the
stem. The terminal spike at the tip of the plant is at least twice
as long as the lateral spikes. The spikes are either reddish or
green. *Amaranthus albus* (tumbleweed) has shorter clusters
of flowers, and the plant tends to be bushy.

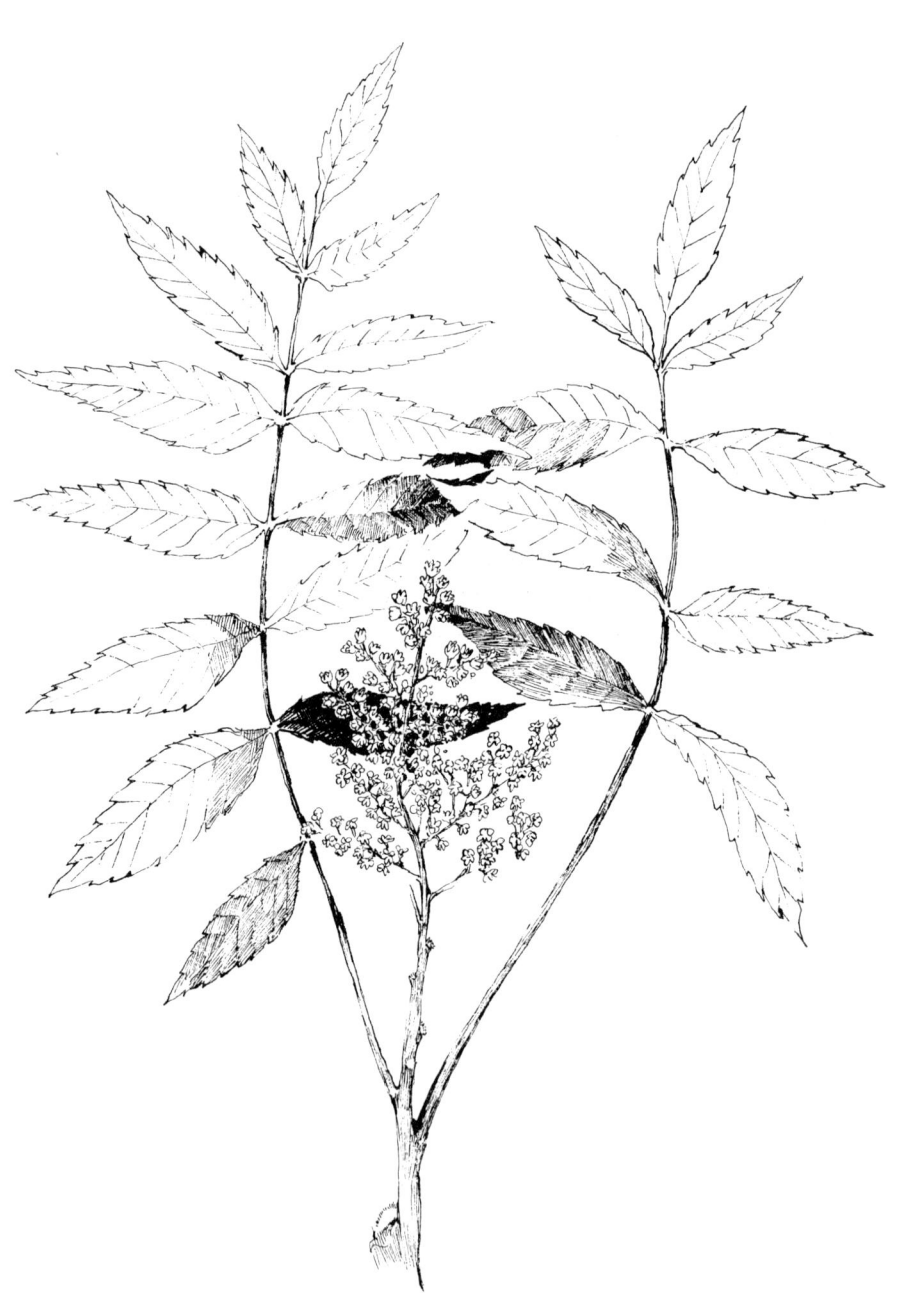

ANACARDIACEAE   *Sumac family*

The sumac family is made up of shrubs or small trees, with many tropical representatives. A few poisonous members are common in Massachusetts. The flowers are small and either unisexual or bisexual, and the leaves are generally compound. The fruit is a drupe.

RHUS GLABRA
Smooth Sumac
Dry fields / June–July / 15 ft.

The pinnately compound leaves of smooth sumac have numerous lance-shaped leaflets which are sharp-toothed and whitish underneath. Its petioles and branches are smooth and it has many, small, greenish-yellow flowers on branched stalks at the tops of branches. The male and female flowers are generally found on separate stalks, although stalks with perfect flowers do occur. The staminate inflorescence is long, while the pistillate inflorescence is smaller and more crowded. The fruits are covered with tiny red hairs. *Rhus typhina* (staghorn sumac) is similar to *R. glabra* except that its branches and leaf petioles are covered with soft, feltlike, light-brown hairs, and the red hairs on its fruits are much longer. Also, the flowers of *R. glabra* are frequently unisexual while those of *R. typhina* are perfect. The wingrib-sumac (*Rhus copallina*) has smooth-margined leaflets, and its branches and petioles have small brown hairs. The stem between the leaflets is prominently winged and the fruits are red and hairy. In Massachusetts, *R. glabra* and *R. copallina* start flowering about one month later than *R. typhina*.

97

RHUS VERNIX
Poison Sumac
Swamps / June–July / 20 ft.

The bark of this sumac is smooth and gray and its leaves are compound with smooth-margined leaflets. The flowers are greenish-yellow and the fruits are grayish-white. The fruits are not clustered at the ends of branches as in other sumacs; instead, they emerge on stalks from along the stem and occur below many of the leaves. Fortunately, this poisonous plant is generally restricted to swampy and somewhat inaccessible areas.

RHUS RADICANS
Poison Ivy
Roadsides, wooded areas | June–July | Vine, ground cover

Poison ivy is an erect, trailing, or climbing shrubby plant with three leaflets which vary widely in form. The flowers are yellow-green and the fruits, grayish-white. Poison ivy is extremely common and frequently is found climbing trunks of trees by way of numerous clinging rootlets. Both poison sumac and poison ivy are dangerous to touch because they will cause a highly unpleasant dermatitis on those who are allergic to these plants.

APOCYNACEAE *Dogbane family*

The members of this family have a milky sap and floral parts grouped in fives: five sepals, five petals, and five stamens. Each flower has two ovaries but only one style and one stigma. The fruit is a follicle.

APOCYNUM ANDROSAEMIFOLIUM
Spreading Dogbane
Fields, roadsides | June-August | 4 ft.

The dogbane is a shrubby plant with paired, egg-shaped leaves along the stem. Its flowers are fragrant and pink with inner red stripes; they generally hang, bell-like, in small groups with the outer lobes of their petals curved backwards. Each of the small scales at the base of the flower tube is opposite a petal lobe. The style is absent and the single stigma sits directly on the two ovaries. The fruits resemble slim pea pods. They grow up to six inches long, and hang downward in pairs since each flower produces two fruits. The seeds have a tuft of hair.

VINCA MINOR
Periwinkle
Roadsides, open woods, escaped from
cultivation | April–May | Ground cover

Periwinkle is used frequently as a ground cover because its stems creep and form mats over the ground. The flower's five petals fuse to form a tube, the inside of which is lined with white hairs. The five stamens have filaments that are fused to the corolla tube for most of their length. The anthers are attached to a tonguelike structure, and the stigma is fringed and has a tuft of white hairs. The style is slender and the two ovaries are bracketed by a pair of nectar glands.

AQUIFOLIACEAE   *Holly family*

This small family of shrubs and trees has mostly unisexual, small, white-greenish flowers borne in the axils of the leaves, either singly or in small clusters. The male and female flowers may be on the same or on different plants. The fruit is a drupe.

ILEX OPACA
American Holly
Woods in coastal regions / June / 40 ft.

The bright red fruits of holly contrast well with its spiny, evergreen leaves. Three other species of *Ilex* occur in the state and all three grow commonly in wet, swampy areas. Their leaves do not have prominent spines. *Ilex glabra* (inkberry), also an evergreen, is coastal in distribution. Its leaves are dotted with glands on their undersurface and they taper near their attachment to the stem. Inkberry has a black fruit. *Ilex verticillata* (black alder; winterberry) has deciduous leaves, elliptical to round, and red to yellow fruits. The sepals of its flowers are fringed with hairs. *Ilex laevigata* (smooth winterberry) is usually found along coastal areas but occurs inland as well. It has deciduous leaves—elliptical to round—and red or yellow fruits. The sepals of its flowers are not hairy.

ARALIACEAE   *Ginseng family*

Representatives of this family in Massachusetts are herbs with
an umbel-type inflorescence, like that of the parsley family.
The flowers are small, white, with five petals, five stamens, one
pistil, and indistinct sepals. The petiole of the leaf does not
sheathe the stem as it does in the parsley family. The fruit is
a berry.

ARALIA NUDICAULIS
Wild Sarsaparilla
Woods / May–June / 12 in.

The leaves and flowers of wild sarsaparilla emerge from a long
rhizome and are borne on separate stalks. The leaves are taller
than the flowers and somewhat umbrellalike. Each plant has
one leaf, divided into three parts, and each part is subdivided
into leaflets. The flower stalk generally bears several umbels.
Aralia hispida (bristly sarsaparilla) grows in woods and along
roadsides and is easily recognized by the numerous bristles on
the lower part of the stem. The umbel is not on a separate
stalk but is borne on top of the leafy stem.

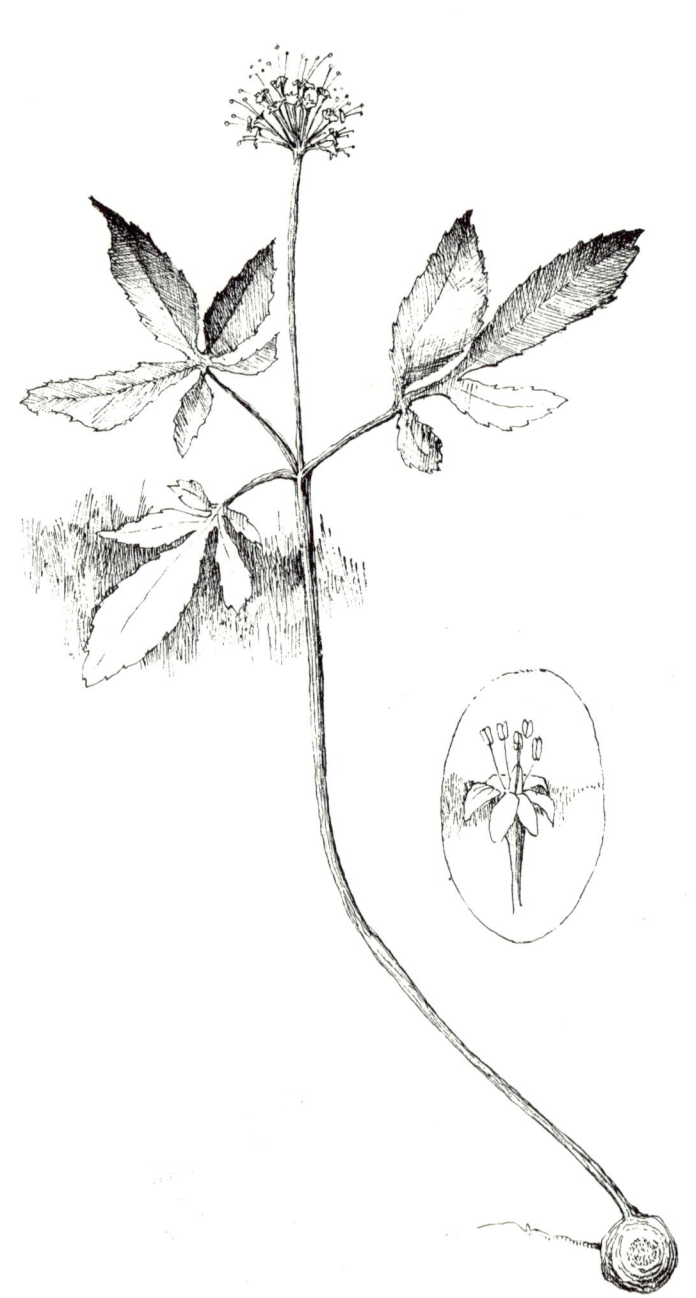

PANAX TRIFOLIUM
Dwarf Ginseng
Woods / May–June / 8 in.

This plant emerges from a deep round tuber from which the stem is easily detached if pulled. The single stem bears a whorl of three leaves, each of which is subdivided into three to five leaflets. Small, white flowers lie in an umbel at the end of a single stalk. The flowers are perfect or unisexual and usually have three or four styles.

ARISTOLOCHIACEAE *Birthwort family*

This is a small family of herbs with flowers at the base of the plant, near the ground. The flowers may be covered with leaves and forest litter and thus may be overlooked. There are no petals, but the flower has fused sepals which resemble petals. The fruit is a capsule.

ASARUM CANADENSE
Wild Ginger
Woods / May–June / 4 in.

Wild ginger has a single, three-lobed, reddish-brown flower at the base of a pair of broad, heart-shaped leaves with long, hairy petioles. Each flower has twelve stamens and one pistil with a six-lobed stigma. The leaves arise from an aromatic rhizome.

ASCLEPIADACEAE *Milkweed family*

Plants of this family are herbs with simple, opposite leaves. A milky sap seeps out when different parts of the plant are cut. The unusual, fragrant flowers are grouped in umbels. Each small flower has five sepals and five petals, all bent backward, with the petals being longer than the sepals. Five cup-shaped, erect structures—actually extensions of the stamens—form a crown in the center of the flower. Each cup has a curved thornlike projection, and each cup, which may contain nectar, surrounds a thick pistil with a thick stigma. The five anthers are fused to the stigma. The pollen grains are fused into a solid mass which resembles a pouch. These pouches occur in pairs and are joined together by a connecting link. One pouch of each pair lies within one anther while the other pouch lies within an adjoining anther. Insects alighting on the flower may tangle their feet on the link connecting the pouches and fly off with the pouches to perhaps pollinate another milkweed flower. The fruit is a large, upright follicle with many parachuted seeds. Only a few flowers in an umbel form fruits.

ASCLEPIAS INCARNATA
Swamp Milkweed
Swamps, wet areas / June–July / 4 ft.

The white to pink flowers of swamp milkweed are borne in several umbels at the top of branched stems. The leaves are lance-shaped.

ASCLEPIAS SYRIACA
Common Milkweed
Fields, roadsides / June-August / 6 ft.

The most common Massachusetts milkweed has thick leaves
that are hairy on their undersurface, and thick unbranched
stems. The umbels occur at the ends of the stems and they
emerge also from the axils of the leaves. The sweet-smelling
flowers are light purple, or sometimes whitish-purple, and the
flower stalks are purplish. *Asclepias exaltata* (poke milkweed)
grows in woods and has drooping greenish-white flowers in
loosely arranged umbels at the top of the plant. *Ascelpias
tuberosa* (butterfly weed) has orange flowers, and it is the
only milkweed without a milky sap. It commonly grows on
Martha's Vineyard and other regions of Cape Cod.

BALSAMINACEAE *Touch-me-not family*

The common name of this family is based on the strange type of fruit its members possess. The fruit is made up of five partitions or valves which separate and recoil quickly when the mature fruit is touched. This sudden reaction propels the seeds out of the capsule. Irregular flowers hang at the ends of stalks and each flower has three sepals. The upper two sepals are small and green; they grip a third, lower sepal which is large and petallike. The large sepal is sac-shaped, open at the front end, and extends into a narrow hooked spur at the rear. Three petals emerge from the mouth of the sac—an upper broad petal and two lateral ones, each with two lobes. Five short stamens with anthers are grouped closely around the stigma.

IMPATIENS CAPENSIS
Jewelweed
Moist, shady areas / July–September / 4 ft.

The flowers of jewelweed are orange with maroon spots. Its stems are smooth and translucent and its leaves have prominent marginal teeth. Large populations of this plant may be present in moist, wooded areas.

BERBERIDACEAE *Barberry family*

This family includes the familiar red-berried barberries, as well as less familiar herbs. Those represented in Massachusetts have flowers with six sepals, six petals, six stamens, and one pistil. The anthers open along their sides by way of little flaps.

BERBERIS THUNBERGII
Japanese Barberry
Roadsides, open fields, woods | April–May | 5 ft.

Commonly used to form hedges, this shrub is now found escaped from cultivation. The small, yellow flowers are borne either singly or in clusters of two to three. Several bracts are located outside of the sepals and each petal has two glands along its base. The fruit is a red berry. The base of the ovary in mature flowers has a large amount of nectar and the single pistil is mushroom-shaped. The stems usually have simple spines, but sometimes lower regions of the plant will have spines with two or three branches. The difference between the Japanese barberry and the common barberry (*Berberis vulgaris*)—the two barberries of this state—is that the Japanese species has leaves with smooth edges while the leaves of the common barberry are toothed. Also, the Japanese species bears its flowers either singly or in small groups whereas the flowers of the common barberry are in racemes.

CAULOPHYLLUM THALICTROIDES
Blue Cohosh
Woods / April–May / 2 ft.

The blue cohosh has one leaf that is sessile on the stem, but because it is divided into three leaflets, each with its own stalk, it appears as if the plant has more than one leaf. Each leaflet is subdivided into three more, lobed units. The lobes of the leaflets resemble the leaves of meadow rue, hence the species name. Another similar but smaller leaf is on the flower stalk. Some parts of the stem are covered with a white bloom. The greenish-yellow to greenish-purple flowers have six, much reduced, fan-shaped petals; these are opposite the sepals and are much smaller in size. The pistil ripens into an unusual fruit which consists of two berrylike blue seeds.

BETULACEAE   *Birch family*

The birch family consists of trees and shrubs with simple, alternate, toothed leaves and reduced flowers arranged in catkins. The male catkins are long and pendulous while the female catkins are short and erect. The fruit is a small nut.

ALNUS RUGOSA
Speckled Alder
Along ponds, streams, and in swamps | Early spring | 12 ft.

The leaves of this shrub are broadest at or below their middle and they have large, sharp teeth along their margins. The female catkins are egg-shaped or somewhat elongated, with overlapping, fleshy bracts. The catkins become woody and conelike and they persist on the shrub after the seeds have become dispersed. The bark has prominent white lenticels. The leaves of *Alnus serrulata* (common or smooth alder) are broadest above their middle and have marginal teeth that are smaller and finer than those of *A. rugosa.*

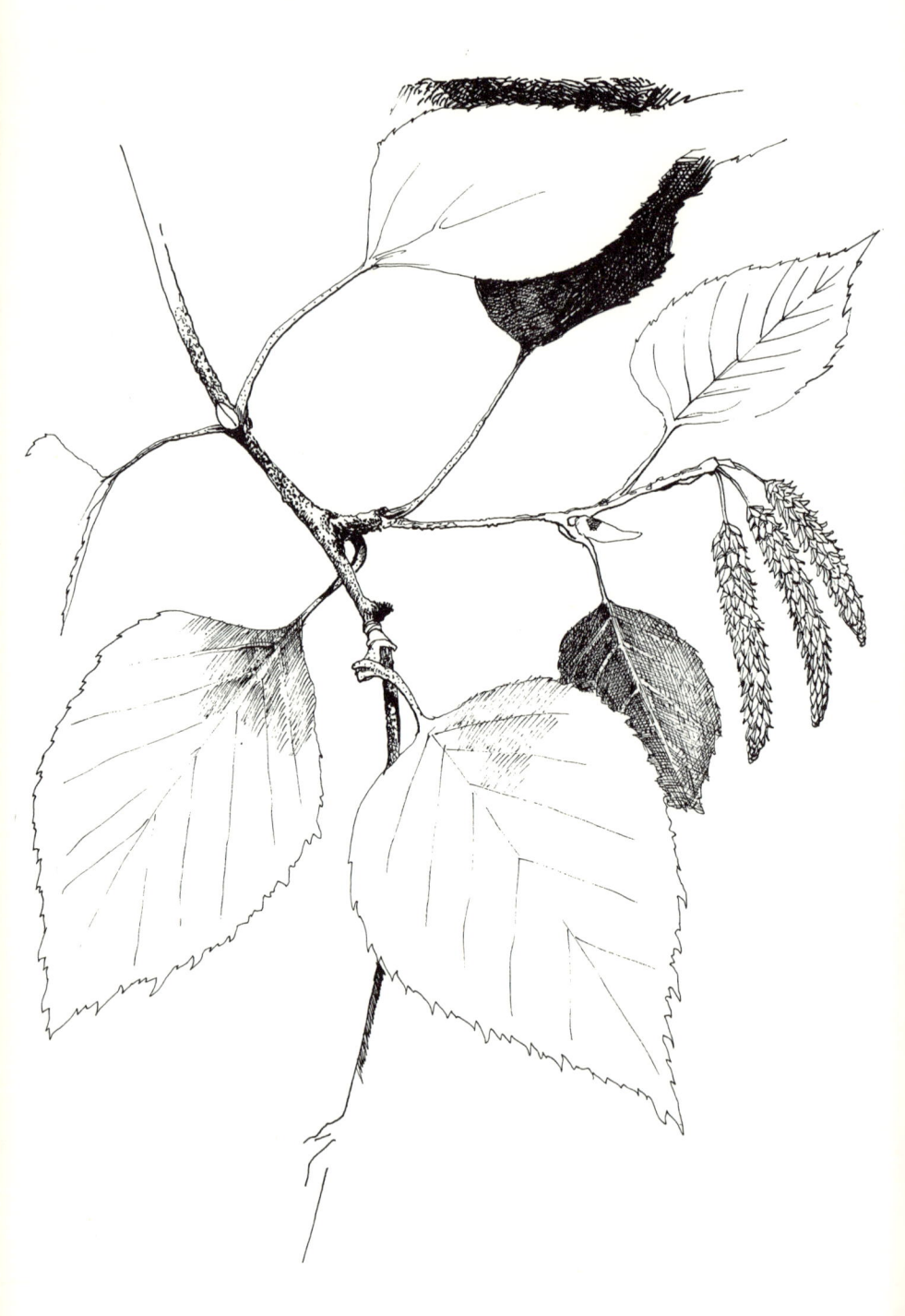

BETULA PAPYRIFERA
Paper Birch
Woods / Early spring / 80 ft.

Paper birch is an easily recognized tree because of its white, chalky bark, which peels away in strips, and its egg-shaped leaves. Other common birches include the following: *Betula lenta* (black or cherry birch) has reddish-brown bark similar to that of a cherry tree; its twigs have a strong wintergreen odor when crushed. *Betula lutea* (yellow birch) has yellowish-gray bark with a sheen; the bark peels off in thin strips and the twigs also have a wintergreen odor when crushed. *Betula populifolia* (gray birch) is generally smaller than the other trees and often more straggly in appearance; its bark is dull white and its leaves are triangular and sharply pointed.

BORAGINACEAE *Forget-me-not family*

This is a family of mostly rough, hairy plants which typically bear their flowers only along one side of stems that unroll as they grow. The fruit is a nutlet.

MYOSOTIS SCORPIOIDES
Forget-me-not
Wet areas, along small streams / June-October / 18 in.

The stems of forget-me-not tend to bend toward the ground, with the tips of their branches curving upward. The stems are angled and branched, and they bear entire, stalkless leaves. The blue flowers occur in rows on the upper ends of the stems. Each flower has five sepals, five basally fused petals, and a raised, yellow center, which is an extension of the corolla tube. There are five stamens, a four-lobed ovary and one simple style.

CAMPANULACEAE   *Bluebell family*

The bluebell family consists of herbs that have flowers with five sepals and five petals. The flowers are bell-shaped in *Campanula* and open in *Specularia*. The five stamens have long anthers and wide, hairy filaments. The stamens are free from each other and are not attached to the petals. The fruit is a capsule.

CAMPANULA RAPUNCULOIDES
Creeping Bellflower
Roadsides / July–August / 3 ft.

This plant spreads by rhizomes. Its blue flowers, an inch or more in length, hang in a row along one side of the upper part of the stem. The style of the pistil is long, hairy, and has three stigmas. Leaves on the upper parts of the stem are lance-like; they are either attached directly to the stem or occur on short petioles. On the lower part of the stem the leaves are heart-shaped and stalked.

CAMPANULA ROTUNDIFOLIA
Harebell
Limestone rock ledges / July–August / 2 ft.

The harebell is found more commonly in the western part of
the state. Its leaves are very narrow except for a few round
basal leaves which disappear after the plant has flowered. The
stems are wiry and branched and the large blue flowers are
borne individually at the ends of branches. *C. aparinoides*
(marsh bellflower) grows in grassy swampy areas and resembles
bedstraw in the way it grows over other plants. It has rough,
three-angled stems and narrow leaves. The single white or pale-
blue flowers occur on long stalks and each flower is about one-
half inch long.

SPECULARIA PERFOLIATA
Venus's looking-glass
Woods / June–August / 2 ft.

This erect plant has an angular unbranched stem and round leaves which lack stalks and clasp the stem. Its flowers also lack stalks and they appear in the axils of the leaves. The flowers on the lower part of the stem are reduced and they never open—they self-pollinate. Flowers on the upper part of the stem do open and have prominent blue petals.

CAPRIFOLIACEAE   *Honeysuckle family*

Members of this family are herbs and shrubs with opposite leaves. The flowers have four to five partially fused petals, inconspicuous sepals, and four to five stamens. Stipules are not present.

DIERVILLA LONICERA
Northern Bush Honeysuckle
Woods / June–August / 3 ft.

This honeysuckle is a shrub with toothed leaves and small groups of funnel-shaped, yellow-green flowers. The five petals of each flower are fused along half their length and the free lobes of four of the petals are rolled back on themselves. The fifth (lowest) petal is not rolled back as much, and it is yellow on its inner side, while the other petals are greenish on the inside. In a group of three flowers, the middle flower does not have a stalk while the two lateral ones do. The flower is hairy on the inside and the fruit is a capsule.

LINNAEA BOREALIS
Twinflower
Woods / June–July / 5 in.

Twinflower is found more commonly in the northern regions of Massachusetts. The main stem trails along the ground, and from that stem emerge erect branches bearing pairs of small, round, evergreen leaves. Each branch ends in a pair of hanging, pink, fragrant flowers. The dry fruit has one seed. The twin-flower was named in honor of the famous Swedish botanist, Carl Linnaeus.

LONICERA MORROWI
Morrow's Honeysuckle
Fields, thickets / May / 10 ft.

This shrub has white to yellowish flowers, widespread petal lobes, a hairy style, and egg-shaped leaves with hairy undersides. *Lonicera canadensis* (fly honeysuckle) grows in woods, is up to six feet tall, and has a smooth style and yellowish flowers. *Lonicera dioica* (wild honeysuckle), a climbing shrub found in woods, has paired leaves joined at their base. In the other two species, the leaves have stalks and are not joined. All three species of honeysuckle produce red berries.

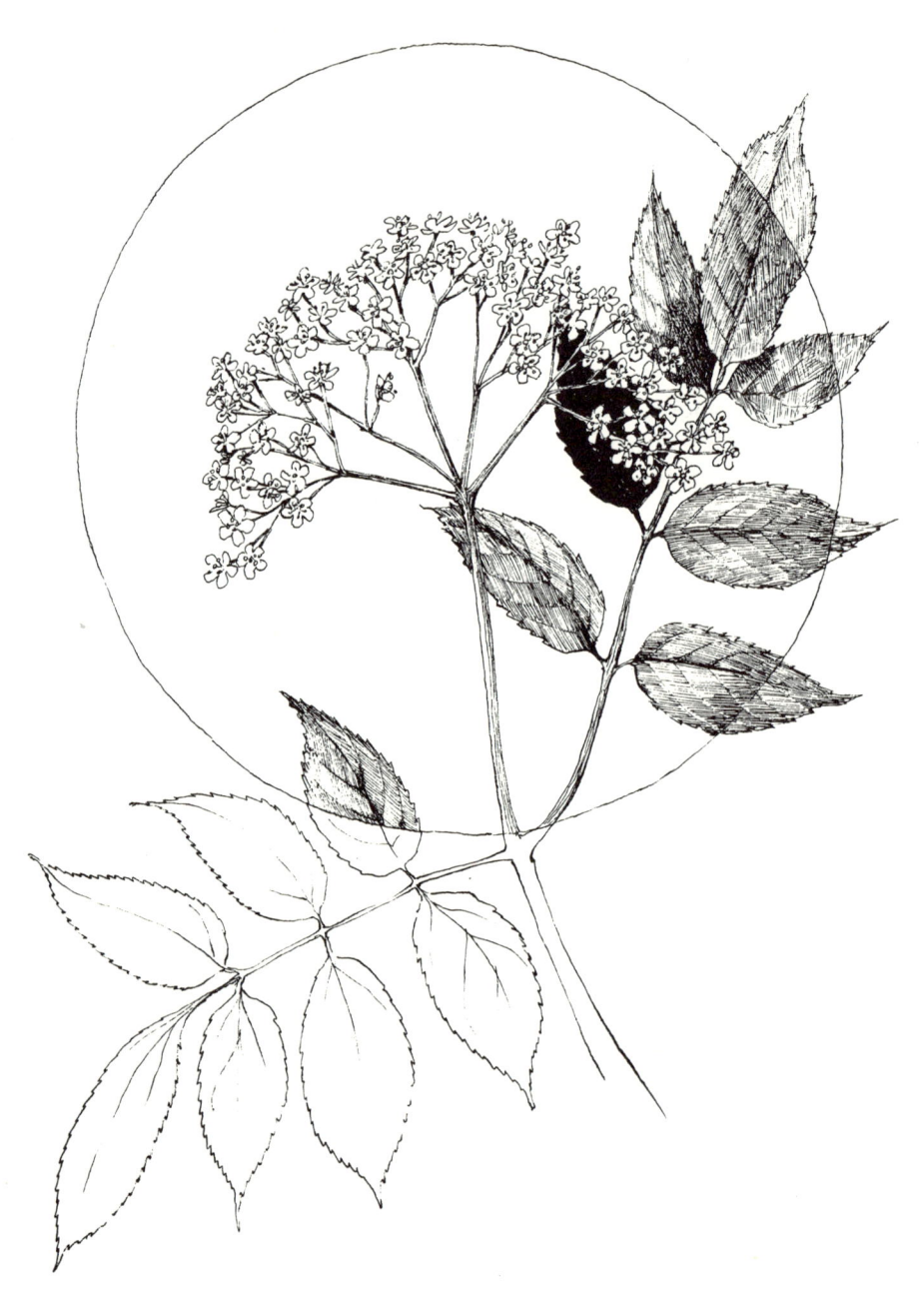

SAMBUCUS CANADENSIS
Common Elder
Woods / June–July / 9 ft.

The common elder is a shrub with pinnately compound leaves
and clusters of small, white flowers in a flat inflorescence. Because the stems have a large amount of white pith, they are
not very woody and thus can bend easily. The leaves have five
to seven saw-toothed leaflets, and the leaf stalks are green. Berries are purple to black and the flower stalks turn purple when
the fruits are maturing. *Sambucus pubens* (red-berried elder)
flowers earlier, in May and early June. Its younger stems and
the underside of its leaves are covered with fine hairs. The inflorescence is cone-shaped rather than flat and the leaf stalks
are purple. It has red berries and a brown pith.

Viburnum differs from *Sambucus* in having simple leaves, whereas *Sambucus* leaves are always pinnately compound. Two groups of *Viburnum* are recognized according to the type of flowers they bear: I. Species with large, sterile, white flowers around a central cluster of much smaller, fertile flowers. II. Species with only small, fertile, white flowers. These plants grow mostly in woods. The fruit is a drupe.

Ia. *Viburnum alnifolium* (6 ft.), hobble-bush, has a sessile inflorescence at the end of each branch, and its leaves are broad, heart-shaped, not lobed, and pinnately veined. Ib. *Viburnum opulus* (15 ft.), high bush cranberry, has three-lobed, maplelike leaves.

IIa. *Viburnum acerifolium* (7 ft.), arrow-wood, also has maplelike leaves. The undersurface of the leaves, the young stems, and the petioles are covered with fine, soft hairs [shown]. IIb. *Viburnum cassinoides* (12 ft.), withe-rod, has simple and unlobed leaves with scalelike particles on the lower surface. Its inflorescence is on a stalk and its leaves have lateral veins that branch and fuse before they reach the margins. IIc. *Viburnum dentatum* (15 ft.), arrow-wood, has leaves with very sharp, large, triangular teeth along their margins, and prominent veins that do not branch more than once or twice, with each vein extending to a tooth. IId. *Viburnum lentago* (18 ft.), nanny-berry, has leaves with small, sharp teeth and winged leaf stalks.

CARYOPHYLLACEAE   *Pink family*

This family of herbs has opposite, simple leaves, and stems that are usually swollen at their nodes. The flowers have five sepals, five petals, ten stamens, and a single pistil with two to five styles. The fruit is a capsule.

LYCHNIS ALBA
White Campion
Along roadsides, waste areas, fields | More common
east of Worcester County | May–September | 4 ft.

Flowers of the white campion have five fused sepals which look inflated. The plants are either male or female; the female flowers have a larger sepal tube than the male flowers. The fragrant flowers open at dusk and close during the day, and each white petal has two lobes. The leaves are hairy and contain glands which produce a sticky substance. *Lychnis dioica* (red campion) is similar but its red flowers open in the morning and close in the evening.

LYCHNIS FLOS-CUCULI
Ragged Robin
Fields and meadows / More common in northeastern
and western Massachusetts / May–July / 2 ft.

The usually pink but sometimes white petals of ragged robin
are each subdivided into four lobes of different lengths, the
two outer lobes being much smaller than the two inner lobes.
The sepals are united into a bell-shaped structure. Another
common name for this plant is cuckoo flower because in
Europe, where it is widespread, the flowers appear at the
time of year when cuckoo bird calls are heard.

SAPONARIA OFFICINALIS
Bouncing Bet
Roadsides / July–September / 3 ft.

Each white or pink-tinged petal of this plant has a small notch, and each petal rolls inward slightly along its margins. The sepals are united into a narrow, vaselike structure. The petals have two parts: an expanded blade that projects beyond the sepal tube, and a long, narrow, tapering stalk that is enveloped within the sepal tube. At the junction of the stalk and blade is a pair of spinelike appendages. The fragrant flowers are often double. The leaves contain soaplike compounds; hence, the generic name.

SILENE CUCUBALUS
Bladder Campion
Roadsides / June–September / 3 ft.

Sepals of the bladder campion unite to form a barrellike structure with a network of prominent veins on its surface. Each white petal is divided into two lobes, and the pistil has three styles. The leaves are lance-shaped with a long, tapering point— they do not have hairs. The stems arise from a rhizome. *Lychnis* resembles *Silene* but differs in having five styles and hairy leaves.

CELASTRACEAE *Staff-tree family*

This family of shrubs has small greenish flowers, simple leaves, and showy, orange-red fruits.

CELASTRUS SCANDENS
Bittersweet
Roadsides, thickets / June / Vine

Bittersweet is a wide-spreading vine with mostly unisexual flowers (i.e., male and female), which occur on separate plants. Some plants may also have bisexual flowers, but they still must be cross-pollinated in order to set fruit. The fruits are widely used for decoration.

CHENOPODIACEAE *Goosefoot family*

The goosefoot family of weedy herbs is similar to the amaranth family but differs in its not having bracts associated with the flowers. The flowers are small and the fruit is one-seeded and indehiscent. Beet and spinach plants belong to this family.

CHENOPODIUM ALBUM
Lamb's Quarters
Waste areas, roadsides / June–September / 3 ft.

Lamb's quarters is a many-branched plant with small, round, greenish flowers. The stems, flowers, and lower leaf surfaces are covered with white, mealy granules. The stems generally are tinged red and the leaves are lancelike with the larger leaves having several coarse teeth on their lower parts. *Chenopodium ambrosioides* (Mexican tea) has an unpleasant smell and leaves that are covered with small, yellow glands. Its lower leaves are cut or toothed.

CISTACEAE *Rockrose family*

Many of the plants in this family commonly grow along coastal, sandy regions. These herbs have simple, alternate leaves, and flowers with five sepals, five petals, many stamens, and one pistil. The fruit is a capsule.

HELIANTHEMUM BICKNELLII
Frostweed
Sandy fields / July–October / 18 in.

The first flowers of frostweed to appear are large with yellow petals, and they occur in small groups. Flowers that appear later are smaller, lack petals, and do not open. These flowers self-pollinate and form fruits. Stems and leaves are hairy. *Helianthemum canadense* (frostweed) is similar but its flowers are borne more or less singly rather than in groups, and they appear a few weeks earlier than the flowers of *H. bicknellii.*

HUDSONIA TOMENTOSA
Beach-heather
Sandy soil along the coast / June–July / 8 in.

The beach heathers are low-growing, bushy plants that form dense mats on the sandy beaches of the sea coast. Their leaves are scalelike and overlapping. Numerous yellow flowers are formed singly at the ends of short lateral branches. *Hudsonia ericoides* (golden-heather) is similar, but its needlelike leaves are erect and do not closely overlap each other.

CLETHRACEAE   *White Alder family*

This family of shrubs has fragrant flowers and simple, alternate leaves with saw-toothed margins. The older bark becomes gray and peels. The fruit is a capsule.

CLETHRA ALNIFOLIA
Sweet or Coast Pepperbush
Wet thickets / August / 6 ft.

These plants produce numerous racemes of very fragrant, white flowers. The regular flower has five sepals, five petals, ten stamens with white filaments, and one pistil with a three-lobed, white style. [shown]

COMPOSITAE   *Daisy family*

The daisy family is one of the largest families of flowering plants with a distinctive inflorescence and many different species. What appears to be the flower of a daisy is not a single flower but rather a close aggregation or composite of numerous small flowers called florets because of their reduced size. The florets are attached to the expanded tip (receptacle) of the flower stalk and they form a type of inflorescence called a head. There are two types of florets. Using the familiar white daisy as an example, one type is the ray floret, which is white, petallike, and radiates from the outer margin of the receptacle. At the base of all the ray florets and enveloping the receptacle are several rows of reduced leaves (bracts) collectively called

163

the involucre. A ray floret pulled from the inflorescence shows a small tubular structure around a single pistil; thus a ray floret is actually a pistillate flower consisting of five fused petals which form a straplike structure and a tube near the point of the flower's attachment on the receptacle. The second type of flower is the disc floret. This is less conspicuous than the ray floret, but there are more of them and they are crowded together on the circular head. The disc florets along the outer part of the head open before those nearer the center of the head. The daisy, therefore, consists of outer white ray florets and many inner, yellow disc florets. The five petals of the disc floret are fused into a tube above which projects a two-lobed stigma. There are five stamens inside the petal tube and their anthers fuse to form a ring through which the style grows. The sepals are either absent or reduced to hairs, bristles, or scales around the base of the petal tube. The fruit is an achene. Of the three types of heads, one type contains only ray florets, as in dandelion and chicory; a second type contains only disc florets, as in thistle; and a third type contains both ray and disc florets, as in the daisy.

ACHILLEA MILLEFOLIUM
Yarrow
Fields, roadsides / June–September / 2 ft.

The yarrow has finely divided, aromatic, fernlike leaves and its stems and leaves have long, white hairs. The numerous heads are small and few-flowered, and together they form a flat inflorescence. The ray florets are generally white, although purplish forms may also be found.

164

AMBROSIA ARTEMISIIFOLIA
Common Ragweed
Fields, waste areas / July–September / 3 ft.

The notorious ragweed is a weedy, hairy plant whose airborne pollen is a common cause of hay fever in the autumn. The leaves are fernlike and the flowers are unisexual. The ragweed's staminate flowers occur in numerous small groups (heads) along long stalks at the ends of branches, while the fewer pistillate flowers occur in small heads below the male flowers. The male flower heads are surrounded by a conspicuous bell-shaped, leafy involucre.

ARCTIUM MINUS
Common Burdock
Roadsides, vacant lots / July–October / 6 ft.

The common burdock has lower leaves which are very broad, coarse, heart-shaped, and easily noticeable even in early spring. The purplish flower heads consist only of disc florets, and each head has a characteristic round involucre made up of many rows of spiny, hooked bracts. The involucre becomes the typical bur of this plant and it surrounds the slender flowers. The flower heads occur singly or in small groups along the stem.

There are about eighteen species of asters found commonly in Massachusetts. The different species are grouped according to the shape of their leaves and to whether or not the leaves are stalked. Two poorly delimited groups of asters are recognized as follows: I. Species whose lower leaves are heart-shaped and have petioles. II. Species whose leaves generally are not heart-shaped and lack petioles, or the leaf narrows near the stem and resembles a broad petiole. Some species have leaves that partially encircle the stem.

Ia. *Aster cordifolius (4 ft.),* Heart-shaped aster, grows in woods; this hairy plant has many small heads on a branched inflorescence, blue, narrow ray florets, and red disc florets. Ib. *Aster divaricatus* (3 ft.), white wood aster, grows in woods; it has crooked stems, leaves, sometimes hairy stems, and white, narrow ray florets [shown]. Ic. *Aster macrophyllus* (5 ft.), wild aster, grows in woods; it has light purple florets, thick leaves, numerous small, glandular hairs on the young stems, and many large, clustered basal leaves. Id. *Aster undulatus* (4 ft.), wavy-leaved aster, grows in woods; it is a hairy plant with light blue rays; its lower leaves are heart-shaped and stalked while the upper leaves are lance-shaped and clasp the stem between their two basal lobes.

IIa. *Aster acuminatus* (3 ft.), whorled aster, grows in woods and has large heads with white rays; its leaves are toothed, broadly lance-shaped, and lack petioles, but they taper toward the stem. IIb. *Aster dumosus* (3 ft.), bush aster, grows in sandy, coastal areas; it has blue rays and many widespread branches; the main leaves near the stem are lance-shaped and sharply toothed, while the leaves on the flower-bearing branches are small and numerous. IIc. *Aster ericoides* (3 ft.), wreath aster, grows in fields and open areas; it has many small heads with white rays and small, narrow leaves that are crowded on the stems and branches; the plant is hairy, and the involucral bracts each have a wide, green spiny tip. IId. *Aster laevis* (3 ft.), smooth aster, grows in open areas and along roadsides; not a hairy plant, it has a grayish bloom over its leaves and stems; the upper leaves are clasping while the lower ones are lance-shaped and may have stalks; the rays are purple. IIe. *Aster lateriflorus* (4 ft.), calico aster, grows in open areas and it has white rays and flower heads arranged in a widely branched pattern; that is, the heads are not clustered tightly together and sometimes occur along only one side of a branch. IIf. *Aster linariifolius* (2 ft.), stiff aster, grows in sandy soil near woods; it is a hairy plant with purple rays and many narrow, almost needlelike, rough leaves. IIg. *Aster novae-angliae* (8 ft.), New England aster, grows in fields and along roadsides; it is a strikingly distinctive plant with purple ray florets and yellow disc florets; it has glandular hairs on the flower stalks and involucre; the stems and leaves are hairy and the leaves are lance-shaped and smooth-margined. IIh. *Aster novi-belgii* (4 ft.), New York aster, grows in swamps and coastal marshes and is more common in the central and eastern parts of the state; it has purple to pink rays and resembles the New England aster except that the New York aster is hairless; the lance-like leaves may have some marginal teeth and the upper leaves are usually much smaller than the lower leaves. IIi. *Aster patens* (4 ft.), late purple aster, grows in woods and has blue rays; it is a hairy plant with heart-shaped leaves which partially encircle the stem and have no marginal teeth. IIj. *Aster pilosus* (5 ft.), white-heath aster, grows in fields and along roadsides; its stems may be covered with white hairs; it has many small, white-rayed heads usually spread along the upper sides of the branches, and its leaves are needlelike. IIk. *Aster puniceus* (8 ft.), swamp aster, grows in swamps and moist fields; it has purple to pink rays, a bristly stem and lancelike, clasping leaves with marginal teeth. Ill. *Aster spectabilis* (3 ft.), showy aster, grows in pine woods along sandy, coastal areas; it has purple rays almost one inch long, large heads, and lower leaves that are lance-shaped, stalked and smooth; the plant is hairy and has glands. IIm. *Aster umbellatus* (6 ft.), flat-topped aster, grows in swamps and wet fields; its flowers have white rays and the heads are clustered at the top of the plant into a flat inflorescence; the leaves are crowded on the stem; they are lance-shaped and without teeth. IIn. *Aster vimineus* (5 ft.), small white aster, is similar to *Aster lateriflorus*; it has purplish stems, white rays, and grows in wet areas.

BIDENS CERNUA
Nodding Beggar-ticks
Wet areas, along ponds | August–September | 3 ft.

The leaves of beggar-ticks are opposite, paired, and lance-shaped and usually have sharp marginal teeth. The leaves do not have stalks and their basal parts tend to wrap around the stem. The flower heads have yellow rays. Two types of bracts make up the involucre: bracts in the outer whorl are large and leafy, while those of the inner set are smaller and scalelike. The flower heads assume a slightly nodding posture as they mature. The fruits (achenes) are flat or four-sided with several barbed spines which enable them to stick to clothing or fur and thus become a great nuisance.

BIDENS FRONDOSA
Common Beggar-ticks
Wet areas, along ponds / August–September / 4 ft.

The leaves of this plant are stalked and pinnately divided into three to five lance-shaped leaflets. The flower heads generally lack rays, or if rays are present, they are yellow and tiny. The outer bracts of the involucre are long and leaflike and they differ in size, while the inner bracts are scalelike. The achenes are barbed.

Several other common species of *Bidens* have been created by professional botanists on the basis of minute differences. Readers who are interested in separating all the species should consult the professional manuals listed under Additional References.

CICHORIUM INTYBUS
Chicory
Fields, roadsides / July–September / 5 ft.

Chicory is a loosely branched and coarse plant with stalkless
leaves whose basal ends partially envelop the stem. The lower
leaves are elongated, pointed, and irregularly cut into lobes
while the upper leaves are small, uncut, and lance-shaped. The
leaves get progressively smaller from the bottom of the stem to
the top. The flowers emerge as pairs or threes, without stalks,
from the axils of small leaves on the upper parts of the stem
and branches. The plant has a long taproot which is the source
of chicory. The flowers close tightly during the afternoon and
open in the morning.

CIRSIUM VULGARE
Bull-thistle
Fields, roadsides / July–September / 6 ft.

The bull-thistle is a large, coarse plant that is covered with yellow-tipped spines. The leaves are pinnately lobed. Where they join the stem, the two edges of the leaves continue downward forming a spiny, winglike extension. The flower heads are borne singly at the ends of the stems or branches and each head has a ball-shaped mass of spiny bracts (the involucre). The individual flowers are tubular (disc florets) and very slender and crowded. *Cirsium arvense*, the Canada thistle, is found in the same habitats and has many pink-purple, small heads with no spines on the involucre.

ERECHTITES HIERACIFOLIA
Fireweed
Marshes, clearings, burned-over areas
August–September / 10 ft.

Fireweed has whitish to light-yellow flower heads with only tightly packed disc florets. Each inflorescence has a swollen base and is surrounded by a row of long, narrow bracts (the involucre). The pappus, or reduced sepals, of each floret consists of many white hairs. With the florets crowded together, the entire mature head has a cottony appearance. Leaves are long, narrow, and sharply toothed and the stem has many fine grooves along its length.

ERIGERON ANNUUS
Daisy Fleabane
Roadsides, waste areas / June–July / 5 ft.

This is a hairy plant whose flowers have white or light purple rays and a yellow center. The inflorescences are smaller than those of robin's plantain and they are more numerous on the stem. The many leaves have sharp teeth; this is especially noticeable on leaves farther down the stem. *Erigeron strigosus* (daisy fleabane) is similar, but the stem has fewer leaves which are generally without teeth. The flower heads of both of these species are about one-half inch wide. *Erigeron pulchellus* (robin's plantain) grows in open fields and along streams. Its flowers appear in May and have lilac or sometimes white ray florets. The flower is about one inch in diameter and the plant grows up to two feet in height. The basal leaves have shallow teeth while the upper leaves are linear and whole. *Erigeron canadensis* (horseweed) can be four feet tall or more; it has many-toothed leaves, a bristly stem, and many small heads on stalks which emerge from the sides of the upper part of the stem.

EUPATORIUM PERFOLIATUM
Boneset
Swamps, along shores / July–September / 5 ft.

Boneset is a hairy plant with coarse, wrinkled, and opposite leaves. The leaves lack petioles and their bases fuse around the stem. The small white flowers are grouped in small heads, consisting of only tubular disc florets and these form a flat-topped inflorescence. *Eupatorium dubium* (joe-pye weed) and *Eupatorium maculatum* (spotted joe-pye weed) have pink-purple flowers and a purplish or spotted stem. The leaves of *E. dubium* are broader than the lance-shaped leaves of *E. maculatum* and they have three main veins, whereas the leaves of *E. maculatum* have one main vein. Both species grow in moist fields and along woodland streams. Also common are *Eupatorium hyssopifolium* (hyssop-leaved thoroughwort), *Eupatorium pilosum* (hairy thoroughwort), and *Eupatorium rugosum* (white snakeroot) all with white flowers. *E. hyssopifolium* grows in sandy soil in fields and open woods and it has very narrow, grasslike leaves. *E. pilosum* has oblong, sessile leaves (two to three times longer than they are broad) and it grows in moist, open areas, and along shores. *E. rugosum* grows in woods and it has broad, egg-shaped leaves with large teeth.

GALINSOGA CILIATA
Quickweed
Roadsides, gardens, waste areas / July–October / 2 ft.

This weedy, hairy plant has small flower heads with white
rays. The heads are borne singly at the ends of long stalks. The
leaves are egg-shaped, opposite, and toothed. The upper leaves
often do not have stalks. The plant generally has a sprawling
habit and is a common weed of potted plants in greenhouses.

HELENIUM NUDIFLORUM
Purple-headed Sneezeweed
Meadows, near rivers / July–September / 3 ft.

The purplish-brown heads of sneezeweed are globelike and
have at their bases yellow, lobed ray florets which hang
downward. The leaves are lance-shaped. They clasp the stem
and extend down it, forming winglike structures. The common name refers to the snufflike properties of the dried flower heads.

Barry Moser

HELIANTHUS DECAPETALUS
Thin-leaved Sunflower
Woods, roadsides / July–September / 5 ft.

The coarse leaves of this sunflower are lance- to egg-shaped, stalked, and they generally occur in pairs. The stem is smooth except near the region of the heads where it is hairy, as are the involucral bracts. The center of the head, consisting of the disc florets, is yellow. *Helianthus divaricatus* (woodland sunflower) has paired leaves, without stalks. The leaves are distinctly lance-shaped and they gradually taper to a long tip. *Helianthus strumosus* (harsh-sunflower) also has stalked leaves, but they are thicker and less conspicuously toothed than those of *H. decapetalus*, and they are light green or whitish on their underside. Both of these species attain a height of five to six feet and they grow in thickets and woods. *Helianthus tuberosus* (Jerusalem artichoke) grows in waste areas and around abandoned dwellings. It grows to ten feet and has tuberous, edible rhizomes, hairy stems, and large lance- to egg-shaped leaves.

Most of the hawkweeds in Massachusetts have yellow flowers, and two groups of these plants are recognized. One group has leaves that are mostly arranged in a basal rosette near the ground. They flower from June to July and they grow in fields and clearings. The second group has, in addition to the basal leaves, a few leaves that are generally much smaller than the basal ones. They flower from August to September, and grow in dry, open woods.

Group I. Hieracium aurantiacum (devil's paintbrush) has orange to reddish flower heads and hairy leaves and stems; the hairs on the middle and upper parts of the stem are black and they are either tipped with a gland or are white with a black base. The stems and leaves of *Hieracium pratense* (meadow hawkweed—shown) are covered with long, brown or black hairs. The hairs are especially common along the leaf margins and veins on the lower leaf surface. The ray florets are yellow and have square tips. *Hieracium floribundum* (smooth hawkweed) has leaves with a smooth upper surface. *Hieracium pilosella* (mouse-ear hawkweed) grows in lawns and usually has only one flower head on a stalk. The undersides of its leaves are hairy and white when young, and the plant produces numerous runners.

Group II. Hieracium gronovii (hairy hawkweed) is a hairy plant with small heads on branches on the upper part of the stem. *Hieracium paniculatum* (panicled hawkweed) is mostly smooth with small flower heads on long, slender and branched horizontal stalks near the top of the plant. *Hieracium scabrum* (rough hawkweed) is a coarse hairy plant with reddish stems; it generally does not have basal leaves.

INULA HELENIUM
Elecampane
Roadsides, fields / July–September / 6 ft.

Elecampane is a plant which has escaped from cultivation. The flower heads are large with narrow, yellow ray florets and yellow disc florets. This is a coarse plant with toothed leaves which are woolly on their undersurface. The upper leaves are stalkless and they clasp the stem, while the lower leaves have stalks.

LEONTODON AUTUMNALIS
Fall Dandelion
Fields, lawns / June–September / 2 ft.

The leaves of this plant are similar to those of the common
dandelion, although they are narrower. The leaves are basal,
tightly grouped, and pinnately compound. The flower heads
occur on separate leafless stalks and each head consists only
of ray florets. The Latin name refers to the prominent toothed
("lion's tooth") leaves, and to the common presence of this
plant in autumn.

RUDBECKIA HIRTA
Black-eyed Susan
Fields, roadsides / June–September / 3 ft.

This familiar plant has hairy, narrow leaves and a single flower head on each stem. The yellow ray florets curve downward while the dark purple disc florets are crowded together to form a conelike structure in the center of the head. Black-eyed Susan and the common white daisy (*Chrysanthemum leucanthemum*) are abundant throughout Massachusetts and are among the most beautiful members of our flora.

SENECIO AUREUS
Groundsel
Wet fields, woods, swampy areas / May–July / 3 ft.

The stems of groundsel are ribbed and its basal leaves, on long stalks, are heart-shaped with rounded teeth. The upper leaves are pinnately divided, mostly without a stalk, attached singly to the stem, and widely separated from each other. The stem generally has numerous small heads with yellow florets. Stems arise from a rhizome either singly or in tufts.

There are about a dozen common species of goldenrods in Massachusetts, all of which add a colorful dimension to the fields and roadsides. They all have yellow rays (except *Solidago bicolor* which has white rays) and they all flower during the summer and fall. The goldenrods are divided into three groups according to how their flower heads are arranged on the plant: I. Plants with many heads that are in small groups at the ends of branches near the top of the stem; the inflorescence has a flat appearance. II. Plants whose flower heads are located in the leaf axils along the stem; that is, not all the heads are concentrated at the top of the plant. III. Plants with small flower heads spread along the upper side of upper branches in dense clusters.

Ia. *Solidago graminifolia* (4 ft.), lance-leaved goldenrod, grows in wet areas; its leaves are narrow, smooth-edged, and fine-tipped, and they have three prominent and parallel veins. [shown on p. 207]

IIa. *Solidago bicolor* (4 ft.), silver-rod, grows in woods; a hairy plant, it is the only common goldenrod with white rays. IIb. *Solidago caesia* (3 ft.), blue-stemmed goldenrod, grows in woods; it has smooth, purplish or bluish stems which often arch toward the ground, and lance-shaped, saw-toothed leaves. IIc. *Solidago puberula* (3 ft.), downy goldenrod, grows in rocky or sandy sites; the plant is covered with sticky hairs and it has narrow leaves; the lower leaves are stalked and toothed while the upper leaves have neither stalks nor teeth.

IIIa. *Solidago altissima* (6 ft.), tall goldenrod, grows in open areas; its stem is covered with fine gray hairs; the upper surface of the leaves feels rough when touched and has fewer hairs than the lower surface. IIIb. *Solidago arguta* (4 ft.), sharp-toothed goldenrod, grows in open areas; its stem is reddish-purple and it has small upper leaves without teeth and broader middle and lower leaves with sharp teeth and fine tips; the stems which carry the flower heads are hairy. IIIc. *Solidago canadensis* (5 ft.), Canada goldenrod, grows along roadsides; it has sharply toothed leaves and it resembles the early goldenrod (*S. juncea*), from which it differs in having hairy stems and leaves; the upper and middle leaves of Canada goldenrod are usually larger and more crowded than the lower basal leaves. IIId. *Solidago gigantea* (8 ft.), late goldenrod, grows in thickets and is one of the tallest of the goldenrods; its stem is smooth and it may have a whitish bloom. IIIe. *Solidago juncea* (4 ft.), early goldenrod, grows along roadsides and in open dry areas; its leaves are narrow and fine-tipped; the lower leaves are large, toothed, and stalked while the upper leaves are small, smooth-margined, and without stalks; the leaves tend to be more widely spaced on the stem than the leaves of Canada goldenrod. IIIf. *Solidago nemoralis* (3 ft.), gray goldenrod, grows in open areas and is one of the shortest of the goldenrods; its stems and leaves appear gray and are covered with fine hairs; the basal leaves are larger than the upper leaves. IIIg. *Solidago rugosa* (5 ft.), wrinkled goldenrod, grows in open areas and is densely hairy; its finely pointed and wrinkled leaves have prominent saw-toothed margins and prominent veins. IIIh. *Solidago sempervirens* (8 ft.), seaside goldenrod, grows along the coast and in salty and brackish water marshes; it has thick stems and smooth-margined fleshy leaves.

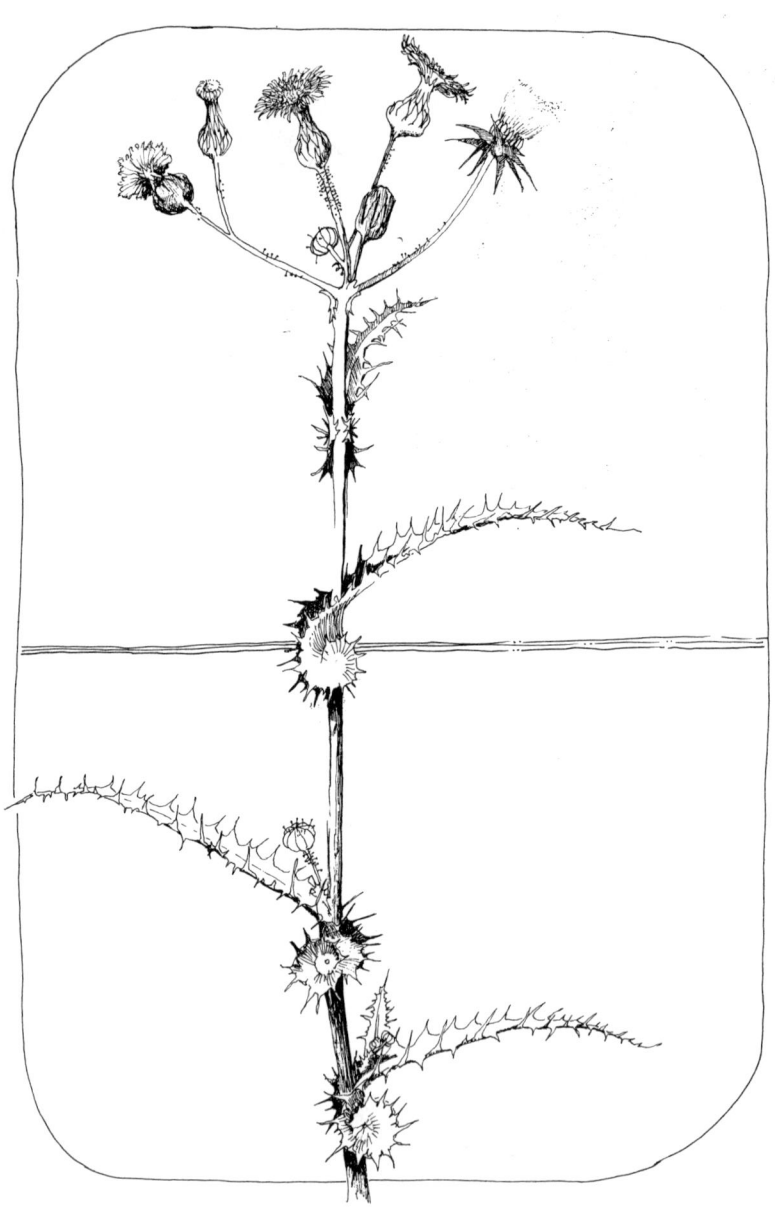

SONCHUS ASPER
Spiny-leaved Sow Thistle
Roadsides, waste areas / July–September / 4 ft.

This sow thistle is a weedy plant with a milky sap and it is easily recognized from its spiny-edged, uncut leaves whose basal parts are curled around the stem. The flower heads consist only of yellow ray florets. *Sonchus arvensis* (field sow thistle) and *Sonchus oleraceus* (common sow thistle) have spiny-edged and generally lobed leaves. The leaves of S. *oleraceus* have bases which envelop the stem and project beyond it, and each base or side of the leaf ends in a sharp point. The leaf bases of S. *arvensis* taper toward the stem but do not project beyond it.

TANACETUM VULGARE
Common Tansy
Roadsides, vacant lots / August–September / 5 ft.

The flower heads of tansy have only yellow disc florets, which are tightly packed in round buttonlike heads that are grouped closely together. The leaves are pinnately divided and very fragrant.

TARAXACUM OFFICINALE
Common Dandelion
Lawns, fields, roadsides / April–August / 18 in.

This familiar weed has a deep taproot and milky sap. Its leaves form a basal rosette and are cut into many toothlike segments. The inflorescence has only yellow ray florets. Around each head there are two rows of bracts, the outer row being shorter and bent backwards. Numerous fruits are produced, each one topped with a hairy parachute. [shown]

TRAGOPOGON PRATENSIS
Goat's Beard
Fields, roadsides / June–August / 2 ft.

The leaves of goat's beard are slender and their bases are wrapped around the smooth stem in a grasslike manner. The flower heads have only yellow ray florets and they close during the day. The fruits taper at their upper end and bear a parachute of bristly hairs. A milky sap is present. This common European plant has firmly established itself in the United States.

TUSSILAGO FARFARA
Coltsfoot
Roadside banks / April–June / 18 in.

This now common European immigrant sends up its flower stalks in early spring, but its broad leaves emerge only after the flowers die. The yellow flowers are packed together in a tight head and each head is at the end of a scaly stalk. The leaves are heart-shaped. They have a deep notch and a lower surface that is covered with white hairs. The plant grows by means of a horizontal rhizome.

CONVOLVULACEAE   *Morning Glory family*

Representatives of this family are vines with a funnel-shaped or tubular flower that is twisted in a bud. Flowers exhibit regular symmetry with five sepals, five petals, five stamens, and one pistil with a two- to three-celled ovary. The fruit is a capsule.

CONVOLVULUS SEPIUM
Hedge Bindweed
Fields, thickets / June–September / 3 ft.

The climbing stems of this plant are frequently entangled among themselves and other vegetation. The leaves are shaped like an arrowhead with the basal part of some leaves spreading out to form two lateral lobes. The flowers are large, white to pink, and trumpet-shaped. The sepals lie closely against each other and they are enveloped by two large bracts. The flowers are usually borne singly, but sometimes several occur together.

CUSCUTA COMPACTA
Dense Dodder
Swamps, meadows / July–September / Vine

Dodder is a much reduced, parasitic plant whose slender orange stems encircle other plants. Small, peglike branches from the stem penetrate the host plants and drain nutrients from them. The leaves are reduced to tiny scales, and numerous bunches of white, glasslike flowers are produced. The flowers are tubular or bell-shaped; opposite every stamen, between the stamen and pistil, is a white, scalelike, erect structure that is fringed or frayed on its upper part. There are two styles, each with a buttonlike stigma. The sepals hug the lower part of the corolla tube. Once the dodder becomes established on a host plant, it loses its connection with the soil and draws all its nutrients from the host. Also common is *Cuscuta gronovii* (common dodder). The distinction between the two is that in *C. compacta* each flower has one or several bracts below it and the sepals are not fused at their base, whereas in *C. gronovii* there are no bracts and the sepals are fused.

CORNACEAE   *Dogwood family*

Members of this family are trees or shrubs—and in one species (*C. canadensis*) an herb—with opposite or alternate, uncut leaves. The flowers are small and grouped together and each flower has four tiny sepals, four petals, and four stamens. Two of the species have showy bracts associated with the flower clusters. The fruit is a drupe.

CORNUS ALTERNIFOLIA
Alternate-leaved Dogwood
Fields, woods / June / 25 ft.

This is the only dogwood with alternate leaves, which are egg-shaped and usually grouped at the tips of the branches; the other shrubby dogwoods have opposite leaves. The fruits are blue. *Cornus stolonifera* (red osier) is common in wet areas and along streams; it has bright red twigs and branches, a white pith, and white fruits. *Cornus amomum* (silky dogwood) grows in similar sites; it has reddish-tinged, hairy twigs, but its pith is brown and its fruits are blue. *Cornus racemosa* (gray-stemmed dogwood) grows in wet wooded areas and along streams; it has a gray bark, a brownish pith, and white fruits. The shrubby dogwoods can be distinguished from shrubs of *Viburnum* by their flowers which have four petals while those of *Viburnum* have five.

CORNUS CANADENSIS
Bunchberry
Woods / May–July / 8 in.

The bunchberry has a group of six leaves near the top of the plant from which arises a single group of tightly clustered flowers. The leaves appear to be in a whorl near the top of the stem, but actually there is a pair of opposite leaves within whose axils emerge four other leaves. The small, greenish-white flowers are surrounded by four large, white, petallike bracts; these give the impression that the whole group of flowers is really one flower. There is a smaller pair of opposite leaves or sometimes scalelike leaves on the stem below the upper whorl of leaves. The fruits are red and bunched together. *Cornus florida* (flowering dogwood) is the other dogwood with showy bracts. Although a variety of forms with pink-red bracts are often seen in cultivated areas, the form with white-bracted flowers is the more common, natural wild type.

CRASSULACEAE   *Live-forever family*

This family of fleshy herbs has simple leaves and bright flowers. The familiar "air plant" (*Kalanchoe*) belongs to this family as do "hen-and-chickens" (*Sempervivum*) and similar cultivated plants.

SEDUM TELEPHIUM
Live-forever
Roadsides, fields / July–September / 2 ft.

This plant has a stout, purple stem and fleshy, green- and purple-tinged leaves with sharp marginal teeth. The rose-colored flowers are crowded together on small branches at the top of the stem and they form one large round inflorescence. Sometimes a plant may have several side branches below the tip, each of which ends in an inflorescence. Each flower has five sepals, five petals, five separate pistils, and eight to twelve stamens.

CRUCIFERAE *Mustard family*

The flowers of this family have a distinctive pattern of four sepals and four petals with the petals arranged in the form of a cross. Each petal narrows into a filament toward its base. Of the six stamens, two are shorter than the others. The one pistil develops into a podlike fruit which is long and narrow in some species and short and round in other species. Fruits are formed in abundance and each fruit has one to several seeds. Many species are troublesome weeds in gardens and lawns.

ALLIARIA OFFICINALIS
Garlic Mustard
Roadsides / April–May / 3 ft.

Both the Latin and common names of this plant refer to the garliclike odor of the crushed leaves. The basal leaves are kidney-shaped with rounded teeth while the upper leaves are triangular and saw-toothed. Flowers are small with four white petals and four boat-shaped sepals which fall off soon after the flower opens. The single, narrow pistil has a head-shaped stigma but no style. Small glands are present at the base of the stamens and the fruits are long and slender.

BARBAREA VULGARIS
Winter Cress
Fields, roadsides | April–May | 3 ft.

Winter cress is a very showy weedy plant because of its numerous, small, bright yellow flowers. The stem is furrowed and the bases of the leaves clasp the stem and partially encircle it. The older leaves are long and subdivided into several leaflets, the terminal leaflet being much larger than the lateral ones. The younger leaves, near the top of the plant, have fewer leaflets.

CAKILE EDENTULA
Sea Rocket
On sandy beaches, along shores / July–August / 12 in.

Sea rocket is a bushy, fleshy plant with unusual fruits. Each fruit has two differently shaped sections: a large upper round part with a pointed tip and a small and narrow lower part. The flowers are small and light purple and the leaves are paddle-shaped and toothed. The lower branches tend to spread out on top of the sand, while the inner branches are more erect.

CAPSELLA BURSA-PASTORIS
Shepherd's Purse
Waste areas, lawns, roadsides, gardens / May–October / 18 in.

Shepherd's purse is a weedy plant with small, white flowers at the top of the stem. Some leaves form a basal rosette while other smaller and stalkless leaves are scattered along the stem. The fruits are distinctive—they are flat and triangular, somewhat heart-shaped, and are borne singly on stalks along the upper half of the stem.

CARDAMINE PRATENSIS
Cuckoo Flower
Meadows / May–June / 2 ft.

The leaves of the cuckoo flower are pinnately compound; the
basal leaves on long stalks are larger than the upper leaves.
The leaflets vary in shape from oval to linear. Flowers are pink
or white. *Cardamine pensylvanica* (common bitter cress) grows
along ponds and brooks and in swamps. It rarely reaches a
height of two feet and it has smaller flowers than those of
C. pratensis. Its leaves also are pinnately divided. *Cardamine
parviflora* (narrow-leaved bitter cress) is a smaller plant than
the other two. Up to eight inches tall, it grows in dry woods
and has small, white flowers and pinnately divided leaves. The
leaflets, however, are short and very slender.

DENTARIA DIPHYLLA
Toothwort
Woods / April / 12 in.

The toothwort grows from a rhizome which has toothlike projections. These projections indicate the points of attachment of aerial stems in past years; i.e., each year the rhizome grows a certain length underground and the following spring sends up a new stem with flowers. The plant has a long-stemmed basal leaf and two stem leaves. Both types of leaves are subdivided into three leaflets with large, rounded teeth. The stem leaves are opposite and the flowers are white.

ERYSIMUM CHEIRANTHOIDES
Wormseed Mustard
Fields, roadsides / May–June / 2 ft.

The leaves and stems of this mustard are covered with small, three-pronged hairs. Leaves are entire and taper toward the base. Small yellow flowers are produced in groups along the upper part of the plant. The fruit pods are long and narrow, four-sided, and more or less erect. They are also hairy and contain many small seeds.

HESPERIS MATRONALIS
Dame's Rocket
Roadsides, woods / May–June / 3 ft.

Dame's rocket has escaped from gardens and it is now growing
wild. It is a tall, hairy plant with showy white to pink flowers.
The sepals are upright, greenish with a white margin, and they
surround the base of the petals. The petals taper toward their
base and form a loose tube around the stamens and pistil. The
flowers are fragrant, especially toward evening.

LEPIDIUM VIRGINICUM
Pepper Grass
Waste areas, roadsides, fields / June–September / 15 in.

This common weed has small white flowers at the tips of branches and numerous, prominent, flat fruits along the upper regions of the branches. Although the usual complement of stamens for the mustard family is six, in this species two or four stamens may not develop. The fruit pod is round and has two halves with one seed in each half. The leaves are toothed, a feature especially noticeable in the basal leaves. *Lepidium campestre* (field cress) has entire leaves whose bases clasp the stem; its fruit pod is longer than it is wide. *Lepidium densiflorum* (prairie-pepperwort) has toothed basal leaves; its flowers are very crowded on the stem, and its petals are either absent or much reduced.

SISYMBRIUM OFFICINALE
Hedge Mustard
Waste areas, roadsides, gardens / June–September / 3 ft.

The small, yellow flowers of hedge mustard are produced in groups at the ends of wide-spreading branches. The fruits are long, slender and upright, and lie against the purplish stem. The upper leaves have a characteristic arrowhead shape with several basal lobes which may arch slightly backward. *Sisymbrium altissimum* (tumbling mustard) has leaves with more lobes, the upper leaves having very thin segments. Also, its fruits are widespread and they are not erect against the stem.

CUCURBITACEAE   *Gourd family*

This family of vines, mostly tropical, has tendrils which help them to climb over other vegetation. The leaves are simple and lobed. The flowers are unisexual; male and female flowers may be on the same or on different plants. There are five to six sepals, five to six petals, and three stamens. The fruit, leathery on the outside and fleshy inside, has few to many seeds. Familiar fruits such as the cucumber, watermelon, squash, and pumpkin belong to this family.

SICYOS ANGULATUS
Bur Cucumber
Along rivers, dumps / August–September / Vine

The flowers of this vine are white or greenish, with the male flowers grouped along erect stalks and the female flowers arranged in small groups on the same plant. The flowers have five sepals and five petals and the fruit is prickly and one-seeded. *Echinocystis lobata* (wild cucumber) is a very aggressive vine and it tends to blanket other vegetation. The white, male flowers are produced on erect stalks which emerge from the axils of maplelike leaves. The flowers have six narrow and tapering, hairy petals, and the sepals are fused into a dish-shaped structure with six bristly projections. The fruit resembles a bladder and is covered with bristles.

247

DROSERACEAE   *Sundew family*

The sundews are insectivorous plants common in bogs and swamps. Their leaves are modified to different shapes and they are covered with sticky hairs on which small insects become trapped and are subsequently digested. In this way, the plants obtain necessary nutrients, especially nitrogen, which are in short supply in the bogs.

DROSERA ROTUNDIFOLIA
Round-leaved Sundew
Bogs, along ponds / June-August / 8 in.

Each plant of sundew has a rosette of small, paddle-shaped leaves and one side of each leaf is covered with reddish hairs. At the tip of each hair is a drop of clear, sticky fluid on which the insects become trapped. Large populations of this plant grow in boggy areas. Its white or pink flowers occur on a long, leafless stalk which emerges from the center of the rosette of leaves. The flower parts are grouped in fives; either the flowers open singly or several open at the same time. Two other species of sundew are found in Massachusetts. One species (*Drosera filiformis*), thread-leaved sundew, grows along the coast on Cape Cod and on Nantucket Island; it has threadlike leaves, up to ten inches long, and purple flowers. The other species (*Drosera intermedia*), spatulate-leaved sundew, is more common along the coast although it grows inland as well. It has oval, spatulalike leaves with long stalks and white flowers.

ERICACEAE *Heath family*

This family includes mostly shrubs—many with evergreen
leaves—and a few herbs. Members of this family grow best in
acid soils and are commonly found in woods, bogs, and on
mountains. The flowers have four to five sepals, four to five
petals, four to ten stamens, and one pistil. In most species the
petals are fused along some of their length and the flowers are
bell-shaped. The anthers open by pores at their tips rather than
splitting lengthwise as in many other families of plants. The
fruit is a capsule or a berry.

ANDROMEDA GLAUCOPHYLLA
Bog Rosemary
Bogs / May–June / 18 in.

The bog rosemary is small and woody with narrow leaves
which are white on their undersurface. The hanging pink flowers are urn-shaped and their component parts occur in groups
of five.

CALLUNA VULGARIS
Heather
Fields / August–September / 18 in.

Heather is a small, branching shrub with tiny overlapping leaves. Small, pink, white, or purplish flowers line the upper ends of the branches. The four, petallike sepals are longer than the four petals. This is a very common plant in Europe. Plant colonies in Massachusetts were probably introduced here by early settlers from Europe.

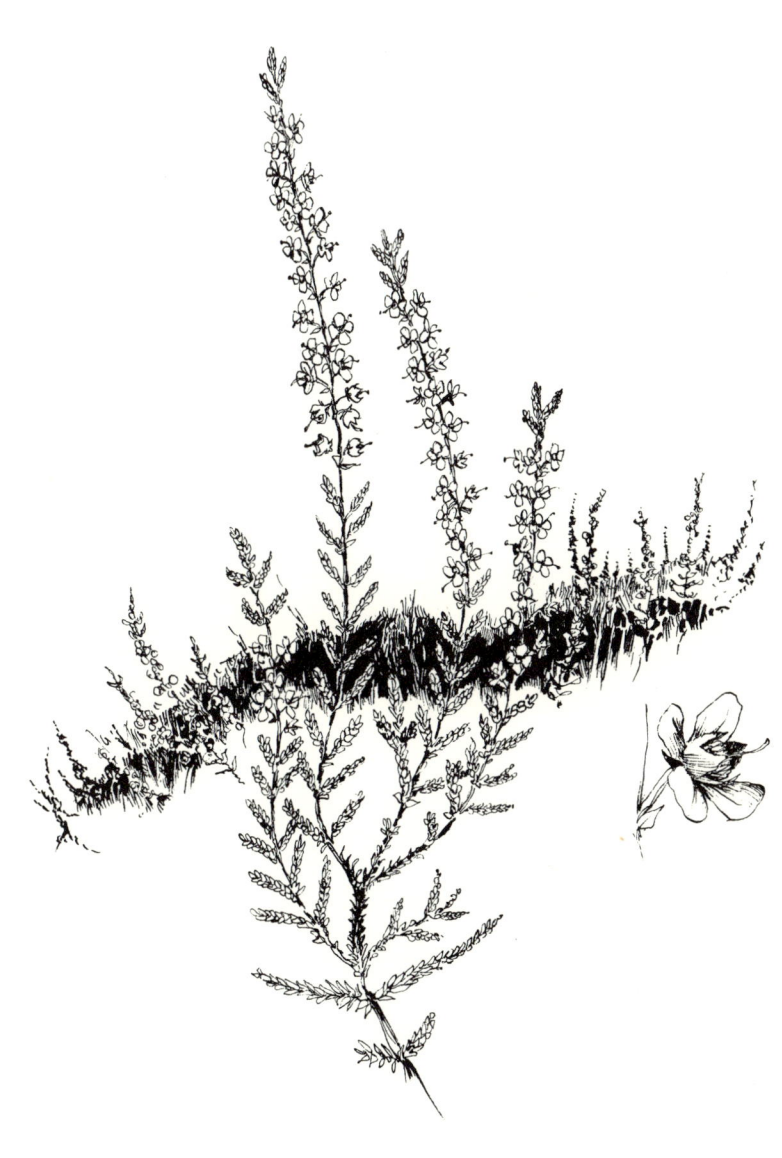

CHAMAEDAPHNE CALYCULATA
Leatherleaf
Bogs / April–June / 4 ft.

Leatherleaf is a common shrub in bogs where it often forms a
dense growth. The bell-shaped flowers have five white petals
whose free ends are yellow, five separate sepals, and two bracts.
There are ten stamens whose filaments are flat, and a short
ovary above a lobed, dark-green disk. The leaves (particularly
the undersurface), the younger parts of the stem, and the sepals are covered with white scales and each scale has a brown
dot at its center.

CHIMAPHILA MACULATA
Spotted Wintergreen
Woods / July / 6 in.

The spotted wintergreen has toothed leaves with pointed tips and white markings. The stem emerges from an underground rhizome and bears one or two whorls of a few dark green leaves. Flowers are on a stalk which extends several inches beyond the upper whorl of leaves. The flowers are sweet-smelling, white, several to a cluster, and they face downward. Each flower has five small sepals, five petals, and a large, flask-shaped, round ovary capped with a disk-shaped stigma. There are ten stamens around the pistil and each anther has a pair of pouchlike chambers (the anther sacs). The anther is attachèd on its side to a wide and hairy filament. *Chimaphila umbellata* (pipsissewa) is found also in woods. It has white or pinkish flowers and toothed, shiny green leaves with broad tips.

EPIGAEA REPENS
Mayflower
Evergreen woods, sandy or acid soils
April–May / Trailing plant

The mayflower is the state flower of Massachusetts and although the plant itself is not prominent, the sweet fragrance of its white to pinkish flowers is distinctive. It is a prostrate, trailing plant with leathery leaves and hairy stems. The flowers have ten stamens and five petals which are fused at their base to form a tube. Around the tube there are five white and green sepals and several keel-shaped bracts. The inner surface of the petals is lined thickly with white hairs.

GAULTHERIA PROCUMBENS
Wintergreen
Woods / July–August / 6 in.

Wintergreen is a low growing plant and it is a common ground cover in woods. The shiny leaves occur on erect stems and the flowers are white, bell-shaped, usually single, and have eight to ten stamens. The flowers hang beneath the fragrant leaves and give rise to spicy and aromatic, red berries. The bright berries may persist on the plant through the winter. In the fall, some of the evergreen leaves turn reddish on plants that grow in exposed areas.

KALMIA LATIFOLIA
Mountain Laurel
Woods / May–July / 10 ft.

The shrubs of mountain laurel are a common sight in the woods of our state. They produce dense, showy masses of pink to white flowers in terminal clusters with each flower borne on a separate stalk. The buds and lower surface of the petals of open flowers are ridged from the pouches that hold the anthers. There are five sepals, five fused petals with ten pouches, ten stamens, and one pistil. The stamens are bent back like catapults and they spring forward when an insect, or finger, touches them. The released anthers strike the intruder and dust it with pollen which is then carried to another flower to bring about cross-pollination. *Kalmia angustifolia* (sheep laurel) is a more slender and upright shrub and it grows in wet, open areas. Its red to purple flowers are not terminal but arise from the axils of the previous year's leaves. The leaves of this plant are reported to be poisonous to livestock. *Kalmia polifolia* (swamp laurel) is a slender, little-branched shrub found in bogs. Its twigs and branches have a crease on each side and its opposite leaves have white undersurfaces. Its red to purple flowers are terminal.

LEDUM GROENLANDICUM
Labrador Tea
Bogs / May–June / 3 ft.

This small, shrubby plant has lance-shaped leaves whose margins are curled under. Young stems and the underside of the leaves are densely covered with a mat of rust-colored hairs. The white flowers are clustered at the top of the plant and they have five petals and five to seven stamens. The crushed leaves are fragrant and, when dried, make a tealike beverage. This is a common shrub in Greenland and Labrador, and on mountains and in bogs farther south.

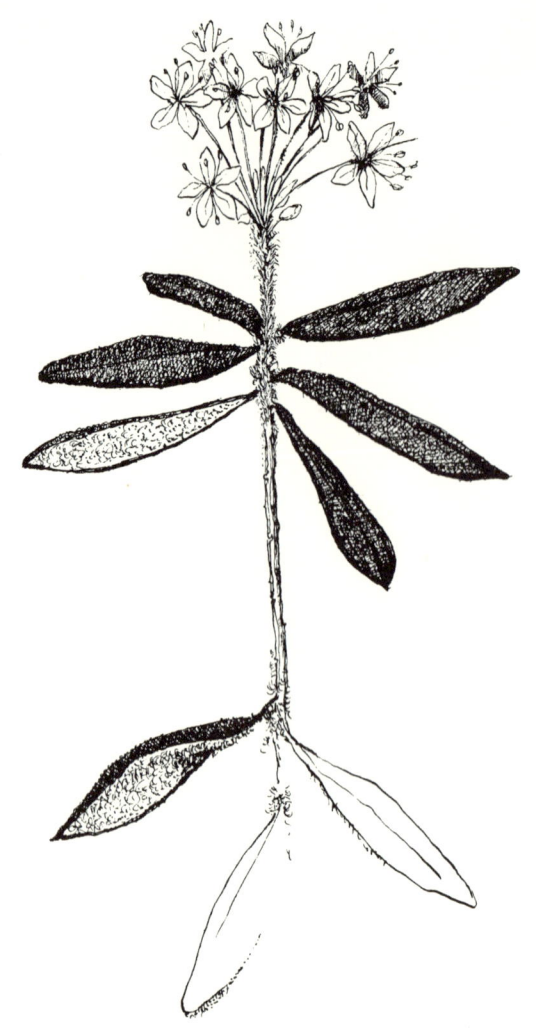

LYONIA LIGUSTRINA
Male-Berry
Swamps, wet woods / June–July / 10 ft.

The male-berry is a tall shrub with entire, uncut, deciduous leaves. The flowers are small, white to pink, and are clustered along a separate stalk which emerges from the point of new growth. The component parts of the flower occur in sets of five. The fruit is a capsule and it splits into five segments. The dry, hard fruits persist on the plant.

MONOTROPA UNIFLORA
Indian Pipe
Woods / June–August / 10 in.

Indian pipe is a white, sometimes pinkish, waxy-looking plant. It lacks chlorophyll and obtains its nutrients from the decaying material of the forest floor. The plant has reduced, scaly leaves and one large, hanging, bell-shaped flower. Each flower usually has five petals and ten stamens; the sepals are absent or fall off early. The ovary is broad, with a short style and broad stigma, and the fruit becomes erect as it matures. Small groups of plants generally grow close together. The plant turns black quickly after it has been picked. *Monotropa hypopithys* (pinesap) has several flowers instead of one and the color of the plant is yellow or pinkish.

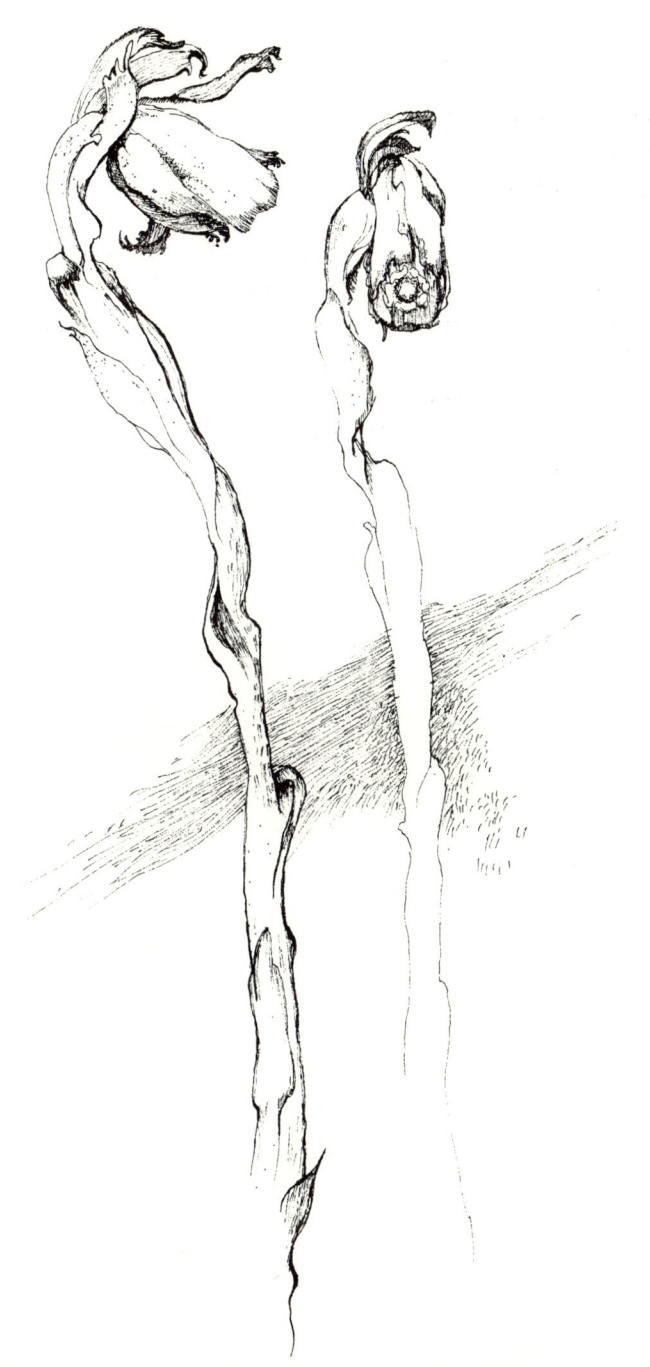

PYROLA ELLIPTICA
Shinleaf
Woods / June–July / 10 in.

The white flowers of shinleaf occur on a long stalk and each flower has five sepals, five petals, a curved style, and ten stamens with yellow anthers. The basal leaves are elliptical to oval. The shinleaf spreads by means of rhizomes. Its fruit is a capsule that opens from the base upward. *Pyrola rotundifolia* (round-leaved pyrola) has round leaves and a leaf stalk that is as long as the leaf itself.

RHODODENDRON NUDIFLORUM
Pinxter-flower
Bogs, woods / May–July / 6 ft.

The pinxter-flower is a deciduous shrub with hairy, white to pink flowers. The flowers either bloom before or at about the same time the leaves emerge. There are scattered hairs on the upper surface of the leaves, but the lower surface is hairless, except along the main vein and leaf margin. Several other common species of *Rhododendron* also are shrubs with deciduous leaves. *Rhododendron roseum* (mountain azalea) is common at high altitudes and is similar to *R. nudiflorum* except that the former has a hairy lower-leaf surface and fragrant flowers. *Rhododendron viscosum* (swamp honeysuckle) grows along shores of ponds, lakes, and in swamps. It blooms from June to July and can grow up to six feet. Its white to pink flowers and flower stalks are covered with red glandular hairs which make the flowers very sticky. The flowers develop after the leaves have grown out. *Rhododendron canadense* (rhodora) grows up to three feet in bogs and woods. It has purplish flowers which appear before or along with the leaves. The flower does not form a distinct tube but rather is split into three parts: an upper, erect, three-lobed part, and two lateral segments.

EUPHORBIACEAE   *Spurge family*

Members of this family are herbs and mostly weeds. Their unisexual flowers are very much reduced, having no petals and
often with no sepals. However, colored bracts may be associated with these inconspicuous flowers. The fruit is a three-part
capsule and each part contains one seed.

ACALYPHA RHOMBOIDEA
Three-seeded Mercury
Fields, roadsides / August–October / 2 ft.

This plant has flowers in small groups at the axils of the leaves.
There are separate clusters of flowers and each cluster has either staminate or pistillate flowers and an underlying bract. The
pistillate bracts are large with five to seven deep lobes, and
around each leaf axis the staminate flowers are borne above
the pistillate flowers.

EUPHORBIA SUPINA
Spurge
Waste areas, roadsides | July–September

This small spurge grows close to the ground and has hairy
stems with usually many branches. Its elliptical leaves may
have purple markings which are small in some plants, but in
other plants they cover most of one side of a leaf. *Euphorbia
cyparissias* (cypress spurge) is common along roadsides, the
plants growing together in large groups. It has needlelike, pale-
green leaves that are often crowded along the stem. The flowers are arranged in an umbel above the plant and each small
group of flowers is associated with a pair of prominent yellowish-green bracts. The plants exude a milky sap when cut, a
characteristic of the genus *Euphorbia*.

FABACEAE   *Bean family*

Members of this family share a **distinctive** type of flower and
fruit. Each flower has five fused sepals and five petals arranged
in a unique pattern. The largest petal, called the standard,
bends upward or stands erect. Two petals, called wings, lie
on each side of the standard; two petals below the wings are
joined along their lower margin to form a keellike structure
within which are the stamens and pistil. There are ten stamens
with fused filaments, or in some species the filament of one
stamen is free from the others. The one pistil develops into a
pod. The leaves are compound and stipules are usually present.

APIOS AMERICANA
Ground Nut
Wet woods near rivers, over stone walls, along
roads through woods | July–August | Vine

The ground nut is a vine with pinnately compound leaves with
five to seven pointed leaflets. Fragrant brownish-purple flowers occur in compact groups along the twining stem. The underground rhizome has several nutlike thickenings.

LOTUS CORNICULATUS
Bird's Foot Trefoil
Roadsides, fields / June–September / 18 in.

This trefoil is a partially prostrate plant. Its leaves are divided
into five leaflets, some of which are egg-shaped while others
taper to a point. Two of the leaflets are close to the stem while
the other three leaflets fan out cloverlike away from the stem.
The flowers are grouped in a circle at the ends of long stalks.
The standard is streaked with red lines and the keel petals
taper to a beak. The slim pods are clustered together and resemble a bird's foot. There are ten stamens—five long and five
short—and the longer ones are enlarged at their tips.

MELILOTUS ALBA
White Sweet Clover
Fields, roadsides / June–September / 9 ft.

Sweet clover is a tall, loosely branched plant with many small, white flowers in long racemes. The leaves are subdivided into three leaflets, like those of clover, and the crushed stems and leaves are fragrant. The flower racemes are best developed near the ends of the lateral branches. *Melilotus officinalis* (yellow sweet clover) grows in similar areas and has the same flowering period. The primary difference between the two is in the flower color. [shown]

ROBINIA PSEUDOACACIA
Black Locust
Fields, roadsides / June / 70 ft.

The hanging white racemes of fragrant flowers and pinnately compound leaves readily identify the black locust. Leaflets are oval and the fruits are hanging pods. Some trees have prominent thorns at the bases of the leaves while others are thornless.

TRIFOLIUM PRATENSE
Red Clover
Fields, roadsides / June-September / 18 in.

The red clover has pink to purple flowers, erect stems, and leaflets with a characteristic v-shaped marking. *Trifolium repens* (white clover) has a trailing stem, white to pinkish flowers, and leaves on separate stalks. *Trifolium hybridum* (alsike clover) has erect stems and a mixture of white to pink flowers on the same head inflorescence. It has lancelike stipules which end in a long, thin tip. The flower heads arise from the leafy stem and they are not on separate stalks. *Trifolium arvense* (rabbit-foot clover) is covered with soft, white hairs. Its leaflets are narrow and the flower heads appear to be gray, although the individual flowers are white-pink. The sepals around each flower are covered with long hairs, which make the heads appear fuzzy. The sepals are fused at their bases, and their upper parts extend into long, toothlike projections above the flower. The individual flowers are hidden among these hairy extensions of the sepals. These sepal extensions are more bristlelike in other clovers. There are two yellow clovers: *Trifolium agrarium* (yellow clover) has an erect habit and the terminal leaflets of its leaf does not have a stalk, whereas *Trifolium procumbens* (low hop clover) has a trailing habit and a distinctly stalked terminal leaflet. The clearest distinction between the two species is that one is erect and the other is prostrate. [*Trifolium procumbens*, shown]

VICIA CRACCA
Cow Vetch
Fields / June–August / 6 ft.

The stems and leaves of cow vetch are covered with soft, white hairs. The leaves are divided into eight to twelve pairs of leaflets; each leaf ends in a tendril which is used to clasp other plants, such as grasses, and grow over them. The numerous blue-purple flowers occur on one side of the flower stalks. The stem is angular and the style is hairy along its upper part.

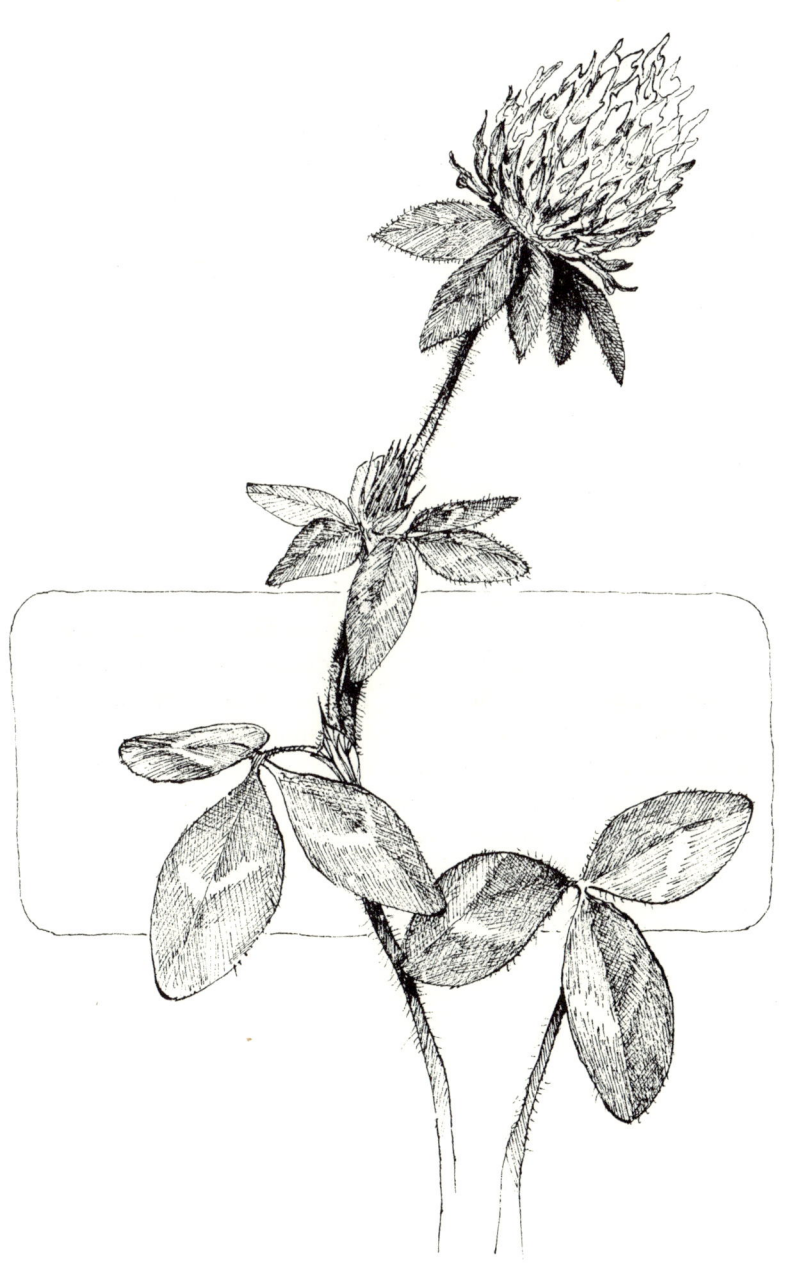

FAGACEAE   *Beech family*

This family of trees has very reduced male and female flowers.
The staminate flowers are clustered on hanging catkins or ball-
shaped heads while the pistillate flowers occur singly or in
small groups. The leaves are simple and alternate and the fruit
is a nut that is enveloped within or partially enclosed by a
coarse or spiny husk called the involucre.

CASTANEA DENTATA
American Chestnut
Dry woods / July / 15 ft.

The American chestnut was once a very abundant and large
tree but now it is severely affected by a fungal blight. New
shoots emerge from old roots and some of the shoots may
reach the size of small trees and produce flowers and fruits.
Flowers appear after the leaves. The leaves are long and narrow, with coarse teeth, and they are smooth on both sides.
Staminate and pistillate catkins occur on the same tree: the
staminate catkins are very long with flowers in small separate
groups and the pistillate flowers are in groups of three within
a circle of prickly bracts (involucre). *Castanea pumila* (chin-
quapin chestnut) is commonly a shrub or sometimes a small
tree with leaves which are hairy on their underside.

FAGUS GRANDIFOLIA
American Beech
Woods / May / 80 ft.

The beech has a distinctive smooth gray bark and long, narrow buds. Its flowers are small and appear along with the leaves. The staminate flowers are on small, hanging, ball-shaped heads while the paired pistillate flowers lie on a short stalk above the staminate heads. The pistillate flowers are surrounded by many awl-shaped bracts. The fruit is a triangular nut that usually occurs in pairs within a prickly husk. The leaves are egg-shaped and have coarse teeth along their margins. *Fagus sylvatica* (European beech) is a commonly planted ornamental tree. It differs from the American beech in having leaves with fewer veins (five to nine *vs.* nine to fourteen) and poorly defined teeth.

QUERCUS ALBA
White Oak
Woods / May / 80 ft.

The white oak has a light-gray bark and its leaves have rounded
lobes. The male flowers are situated on hanging catkins which
appear along with the leaves, and the female flowers occur
singly or in small groups. The fruit is an acorn. *Quercus bicolor*
(swamp white oak) and *Quercus prinus* (chestnut oak) do not
have lobes on their leaves, only rounded, shallow teeth along
the leaf margins. The other oaks included here have distinctly
lobed leaves. While *Q. bicolor* has leaves which are gray and
hairy underneath, the lower surface of *Q. prinus* leaves are
green and usually only sparsely hairy. *Q. borealis* (red oak) and
Q. velutina (black oak) have a stiff hair at the end of each leaf
lobe. This hair is not present in the other oaks. *Q. velutina* has
hairs on the lower surface of its leaves while the undersides of
the leaves of *Q. borealis* are hairless.

FUMARIACEAE   *Bleeding-Heart family*

This small family of herbs has compound leaves and irregular flowers. Some of the petals form a sac or pouch. There are two sepals, two outer and two inner petals, six stamens, and one pistil. The fruit is a capsule.

DICENTRA CUCULLARIA
Dutchman's Breeches
Woods | April–May | 12 in.

The fragrant flowers and finely divided leaves of dutchman's breeches occur on separate stalks which grow out from small, underground tubers. The two outer petals form inflated pouches and each arching flower stalk bears a row of several to perhaps a dozen white with yellow-tipped flowers.

GENTIANACEAE *Gentian family*

This family of herbs has four to five petals, sepals, and stamens. The petals are fused along most or some of their length and the fruit is a capsule.

BARTONIA VIRGINICA
Bartonia
Bogs, wet fields / August–October / 12 in.

The leaves of bartonia are tiny and often barely visible with the naked eye. They are spaced along a thin, yellow stem on the upper part of which emerge stalks with small yellow flowers.

GENTIANA CLAUSA
Closed Gentian
Along rivers, wet fields / August–October / 2 ft.

This unusual plant has flowers which always remain closed.
Between the lobes of the five petals and joining them in pairs
are folded extensions or plaits of the petals. The plaits are
slightly shorter than the petal lobes and are jagged at their tips.
Small groups of flowers are produced in tight clusters at the
top of the stem; in well-developed plants, flowers are produced
also from the lower leaf axils. The paired leaves are lance-
shaped, simple, and generally without stalks.

GENTIANA CRINITA
Fringed Gentian
Along rivers, wet fields / August–October / 2 ft.

The large, bright blue flowers of the fringed gentian are borne singly on long stalks. Each flower has four petals that are finely tattered along their outer margins. There are no folds or plaits between the lobes of the petals. The leaves are lance-shaped, simple, in pairs along the stem, and without stalks.

MENYANTHES TRIFOLIATA
Buckbean
Bogs / May–June / 12 in.

The flowers and leaves of buckbean occur on separate stalks which grow out from a long rhizome situated under water. The flowers are grouped along the upper end of the stalk, and flower parts are in groups of five. The white to pinkish petals are covered with hairs on their inner surface. The leaves are compound with three leaflets. Two types of flowers are produced: one type has a long pistil and short stamens; another type has a short pistil and long stamens.

GERANIACEAE   *Geranium family*

The flowers of this family of herbs have five sepals, five petals, ten stamens, and a pistil made up of five fused carpels with curled stigmas. The carpels each contain one seed and they separate from their base and curl upward when mature. The style of each carpel elongates and the five fused and elongated carpels resemble a bird's bill—from whence comes the common name of cranesbill for some members of this family. The leaves are mostly opposite and have stipules at the base of their stalks.

GERANIUM MACULATUM
Wild Geranium
Woods, roadsides, fields / May–June / 2 ft.

The wild geranium has pink to purple, hairy flowers in groups of two to five. The leaves are subdivided into five to seven lobes; some leaves are basal with long stalks while other leaves are attached to the upper part of the stem.

HAMAMELIDACEAE   *Witch Hazel family*

This family of woody plants has simple, alternate leaves and flower parts in sets of four. The fruit is a capsule which becomes woody as it matures.

HAMAMELIS VIRGINIANA
Witch Hazel
Woods / September–November / 15 ft.

The witch hazel is a shrub or small tree with several interesting features. It blooms late in the season and its seeds are dispersed "explosively" from the fruits to a distance of over ten feet. The flowers have yellow, threadlike petals and the leaves are broad, with rounded teeth and unequal bases.

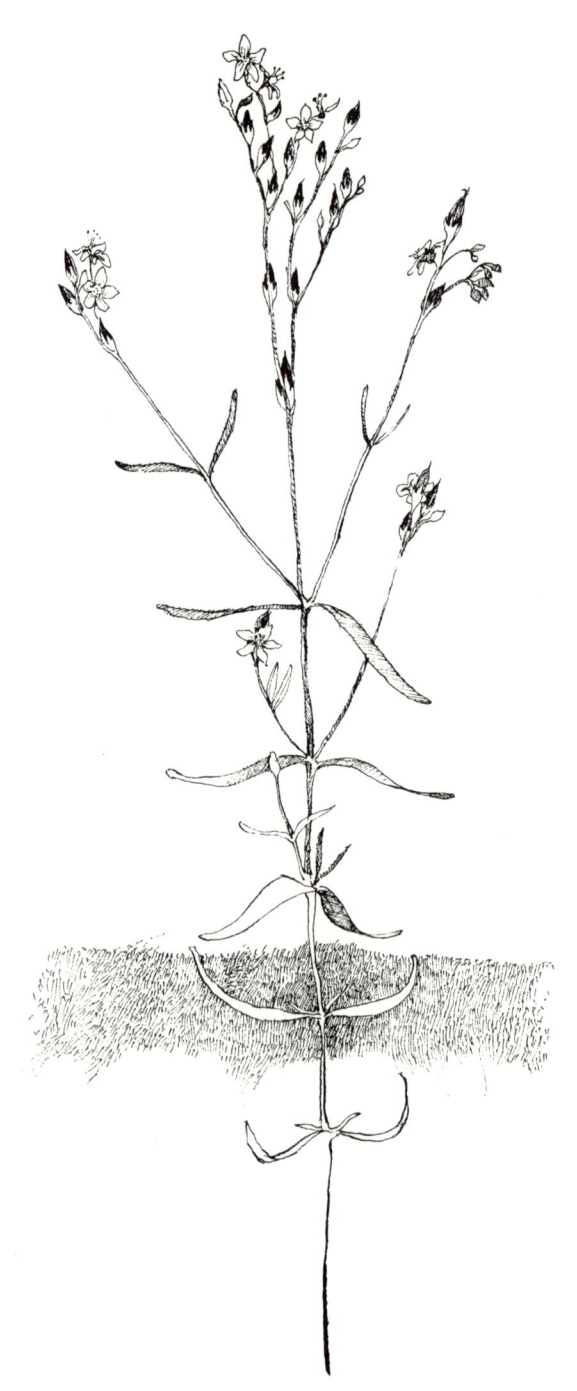

HYPERICACEAE *St. John's Wort family*

A distinguishing feature of this family of herbs is the numerous glands that dot the leaves and flowers. The glands can be seen as conspicuous black specks on the leaves, stems, or petals or as small, clear spots within the leaves. The yellow flowers have five sepals, five petals, many stamens, and one pistil with several widespread styles. The leaves are simple and opposite and the fruit is a capsule.

HYPERICUM CANADENSE
Narrow-leaved St. John's Wort
Bogs, swamps / June–August / 18 in.

The small flowers of this plant do not have any black dots and its leaves are narrow and linear.

HYPERICUM PERFORATUM
Common St. John's Wort
Roadsides, dry areas, fields / June–August / 2 ft.

Many stamens radiate from the flower of the common St. John's wort. The pistil has three separate styles and a round, three-part ovary. The petal margins have black glands. The leaves are studded with clear dots on their upper surfaces and they have scattered black dots on their undersides. Black glands are present also on the stem, especially along the crease which runs on both sides of the stem. The stamens are joined at their bases into several groups, a trait found also in *H. punctatum* but not in the other species. *Hypericum boreale* (northern St. John's wort) grows in bogs and swamps and it has egg-shaped leaves and a branched stem with small flowers. *Hypericum ellipticum* (pale St. John's wort) grows in swampy areas and along streams and it has elliptical leaves and a few flowers on top of an unbranched stem. *Hypericum gentianoides* (pine-weed) grows in sandy soils; it has very small flowers, one at the end of each of many branches, and it has tiny scalelike leaves. *Hypericum majus* (greater St. John's wort) grows in wet fields and along streams and has lance-shaped leaves whose bases meet around the stem. *Hypericum punctatum* (spotted St. John's wort) grows along streams and over wet areas; its leaves and flowers are heavily spotted with black glands.

JUGLANDACEAE   *Walnut family*

This family of trees has pinnately compound, alternate leaves. The staminate and pistillate flowers occur in separate groups on the same tree. Staminate flowers are in hanging catkins while the female catkins are in shorter, erect spikes. The fruit is a nut enclosed within a husk.

JUGLANS CINEREA
Butternut
Woods / May / 80 ft.

Each leaf of the butternut is subdivided into seven to seventeen leaflets which taper to a long point. The bark is gray and the pith of the branches is a dark brown. *Carya ovata* (shagbark hickory) has five to seven leaflets in each leaf and it is easily identified by the way the bark peels from its trunk.

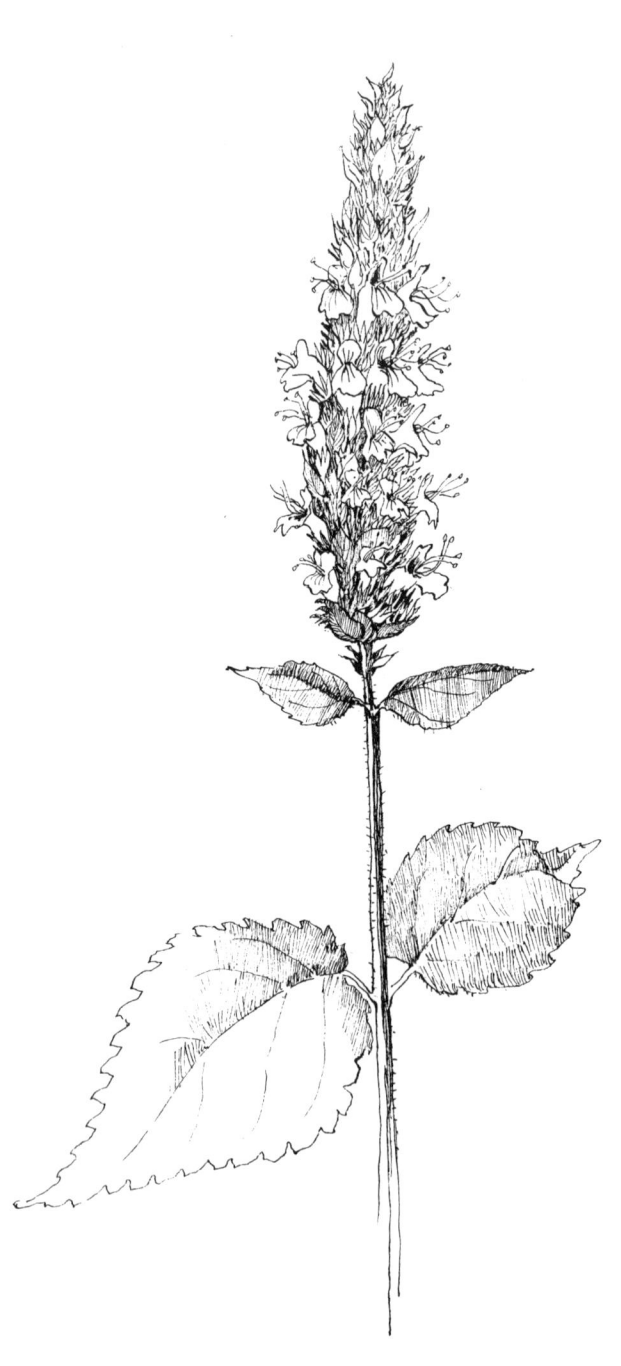

LABIATAE   *Mint family*

One of the easiest ways to assign unknown plants to this family of aromatic herbs is to roll the stem between thumb and forefinger. If the stem is square, the plant most likely belongs to the mint family. With the sedges (Cyperaceae), a similar test will reveal a triangular stem. Other characteristics of the mint family are simple, opposite leaves and an irregular flower made up of five sepals and five petals. In most species, the petals fuse to form a tube with an upper and a lower lip. The sepals also may fuse to form a tube. There are two to four stamens; they are either equal in length, or there are two long and two short stamens. The filaments of the stamens are fused to the corolla tube. The ovary has four lobes and each lobe forms a small nutlet.

AGASTACHE SCROPHULARIAEFOLIA
Giant Hyssop
Fields, open woods | June–August | 4 ft.

The purple flowers and associated bracts of this somewhat rare plant are clustered along a spike at the top of the stem. The stem and underside of the leaves are hairy and the leaf margins have prominent teeth. Four stamens protrude beyond the two-lipped flower. The two lower stamens curve upward and cross the two upper stamens which bend downward.

COLLINSONIA CANADENSIS
Horse Balm
Woods / July–September / 3 ft.

The leaves of horse balm are large and egg-shaped and have prominent, sharp teeth along their margins. The fragrant yellow flowers are grouped in racemes along branches which stand above the leaves. The lower lip of each flower is fringed and extends beyond the upper lip. There are two long stamens and one equally long pistil.

GALEOPSIS TETRAHIT
Hemp Nettle
Roadsides, fields, woods / July–September / 2 ft.

Glandular hairs and bristles cover the stem, corolla tube, and upper lip of the hemp nettle's flower. The corolla tube is slender and it flares into two lips. The middle lobe of the lower lip has two ridges on either side near the mouth of the corolla tube. The white to purple-tinged flowers have four stamens, two of which are slightly longer than the other two. The sepal tube is also bristly and it ends in five long spines. The stems are swollen at the nodes, where the branches and leaves emerge.

GLECHOMA HEDERACEA
Ground Ivy
Roadsides, yards, woods | April–June | Ground cover

This plant has a creeping habit but it sends up numerous, erect stalks with small, rounded, fanlike leaves. Blue flowers, usually in clusters of three, emerge from the leaf axils but all the flowers do not mature at the same time. Five sepals fuse for most of their length to form a tube within which the corolla tube emerges. The lower lip of the corolla tube has three parts, the large middle part of which is spotted and bearded. The upper lip of the flower is small and has two parts. There are four stamens—two long and two short—and one long pistil.

HEDEOMA PULEGIOIDES
Pennyroyal
Fields / July–September / 12 in.

Pennyroyal is a strongly aromatic and hairy plant with small
blue flowers which are clustered around the axils of the leaves
along the length of much of the stem and branches. The sepal
tube is conspicuously ribbed and has five upward-turning
spines: three spines are on the upper lip and two spines are on
the lower lip. There are two stamens and the leaves are lance-
shaped or oval.

LAMIUM AMPLEXICAULE
Dead Nettle
Roadsides, waste areas / May–September / 12 in.

The stems of dead nettle are upright, or they arch; the erect stems bear the leaves and flowers. The few basal leaves are on long stalks while the upper leaves are without stalks. Margins of the small leaves have rounded teeth. The purple, spotted flowers occur in whorls around the leaf axils and they have two long and two short stamens.

LEONURUS CARDIACA
Motherwort
Waste areas, roadsides / July–August / 5 ft.

Motherwort has purple tinged stems except near the tips where they are green. Along the stem there are many long-stalked leaves with three or more finely pointed lobes. Small, purple flowers form clusters in the leaf axils and because the leaves are opposite, the flowers appear to be whorled around the stem. The calyx tube has five spiny projections: three are erect and two bend backwards. The upper lip of the corolla is covered with white, soft hairs. There are four stamens of equal length. [shown]

LYCOPUS AMERICANUS
Bugle-weed
Swampy areas, wet woods, thickets / July–September / 3 ft.

Another common name for this plant is cut-leaved water-horehound. Its leaves are lancelike and have prominent teeth or even small lobes along their margins. Tiny, white flowers are clustered tightly around the leaf axils; these groups of flowers are tiered all along the stem. There are two fertile stamens and sometimes two additional reduced and nonfertile stamens.

MENTHA ARVENSIS
Mint
Wet areas / July–September / 2 ft.

The small, purple flowers of mint are crowded around the leaf
axils along the upper parts of the stem. The leaves are lance-
shaped or egg-shaped and have a strong, minty aroma. The
leaves and stems are hairy and the stems are more or less erect.
There are four stamens of equal length. Cultivated mints commonly escape from gardens and establish themselves along
roadsides.

MONARDA FISTULOSA
Wild Bergamot
Roadsides, woods / July–August / 2 ft.

Wild bergamot is a showy, hairy plant with light purple flowers bunched together at the ends of the stems and branches. The upper lip of each flower is long, narrow, and tubular while the lower lip is broader and curves downward. The flowers have two stamens, and the base of each flower cluster is surrounded by wide, purplish bracts. Leaves are toothed and lance-shaped. *Monarda media* (purple bergamot) has reddish-purple flowers and brownish-green leaves. [shown]

NEPETA CATARIA
Catnip
Waste areas, fields, roadsides / July–September / 3 ft.

The stems and leaves of catnip are covered densely with short hairs, and the entire plant appears grayish-white. The leaves are more or less egg-shaped and have large, coarse teeth. Flowers are white with pink spots and they are thickly grouped at the ends of the stems and branches. Sometimes, smaller groups of flowers emerge from the leaf axils below the main, terminal inflorescence. There are two long and two short stamens.

PHYSOSTEGIA VIRGINIANA
Obedient Plant
Roadsides, fields / July–September / 3 ft.

The common name of this plant originates from the ability of its flowers to swivel from side to side. If a flower is pushed to one side, it remains there and therefore is "obedient." The purplish-white flowers are crowded in spikes along the upper parts of the plant. Smaller spikes grow out from the axils of some of the upper leaves. The upper lip of the corolla is single and uncut while the lower lip has three lobes. The inside of the flower has purple dots and lines. There are four stamens, two of which are slightly shorter than the other pair; the hairy filaments of the stamens are fused to the corolla tube. The leaves are lance-shaped with prominent, sharp, marginal teeth.

PRUNELLA VULGARIS
Self-Heal
Fields, roadsides / June–September / 18 in.

Self-heal is a creeping or erect plant with simple, lance-shaped, entire leaves. Its purple flowers are grouped in spikes at the tops of the stems. The upper lip of the corolla tube is hood-shaped and its fringed lower lip has two small, lateral lobes. The calyx tube is two-lipped and hairy: its lower lip is cut into two sharp lobes and its upper lip is broad and larger than the lower lip. Wide bracts which resemble sepals occur at the base of small groups of flowers. Each flower has two long and two short stamens. The filament of each stamen is forked at its tip and the anther is borne on one of the forked tips.

PYCNANTHEMUM MUTICUM
Mountain Mint
Woods / July–September / 2½ ft.

The mountain mint has small, white flowers which are crowded together, along with their associated bracts, at the ends of branches along the upper part of the plant; or the flowers may occur in the leaf axils. Each group of flowers resembles the head inflorescence of a daisy. There are four stamens. The leaves are egg-shaped to lance-shaped and the stem and small leaves below each flower cluster are hairy, as are the sepals and bracts. The upper leaves and bracts are grayish-white. Several other species of mountain mint with narrow, needlelike leaves have been recognized.

SATUREJA VULGARIS
Wild Basil
Woods, clearings / June–August / 18 in.

Wild basil is a hairy plant with usually unbranched, erect stems. The reddish-purple flowers occur in small groups at the ends of the stems or branches. The calyx tube has spiny projections. There are two long and two short stamens, and the leaves are more or less egg-shaped.

SCUTELLARIA LATERIFLORA
Mad-Dog Skullcap
Meadows, wet woods / July–August / 2 ft.

The flowers of this plant are clustered along racemes which emerge from leaf axils along the upper half of the stems. Purple to pink flowers occur in pairs on one side of the racemes. The sepals are fused into a two-lipped tube which has a small cap on its upper half. The hooded, upper lip of the corolla tube arches over the spreading lower lip. Two, small lateral lobes are mostly tucked into the upper lobe. The pistil and four stamens are located under the upper lip, and the inner surface of the lower lip is whitish and speckled. Leaves are stalked and lance- to egg-shaped. *Scutellaria epilobiifolia* (common or marsh skullcap) has single purple flowers in the axils of the leaves and all the flowers occur on the same side of the stem. The leaves are mostly without stalks.

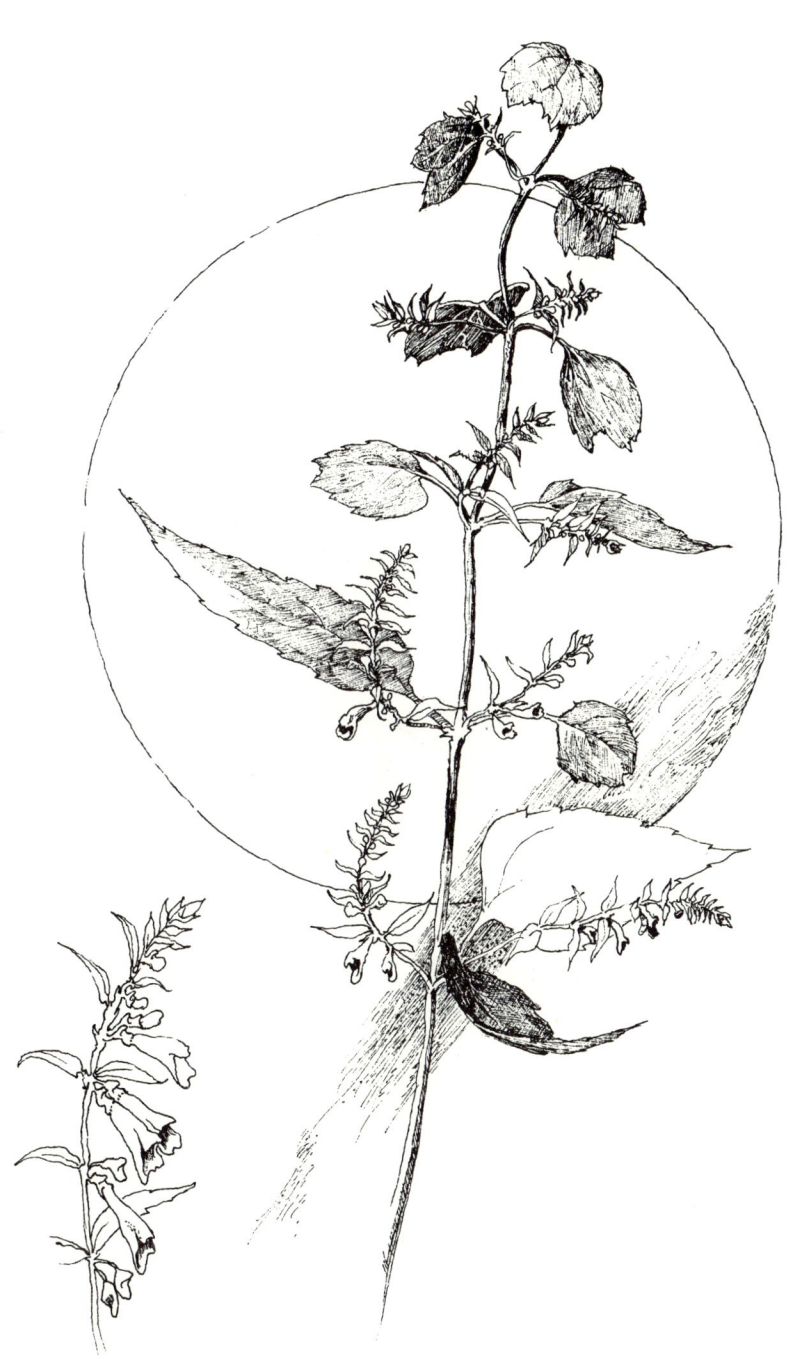

STACHYS PALUSTRIS
Hedge Nettle
Wet fields, along shores / June–September / 3 ft.

Hedge nettle is a hairy plant with lance-shaped, toothed leaves.
The leaves are large basally and they become smaller toward
the upper part of the stem. The purplish, spotted flowers are
found in small groups along the top of the stem; each group
of flowers lies in the axils of a pair of leaves. The sepal tube
is hairy and has spiny lobes. *Stachys hyssopifolia* (hyssop
hedge nettle) grows along the shores of ponds in sandy areas
near the coast. It is a smaller plant—up to eighteen inches—
hairless, and has slender, smooth-margined leaves.

TEUCRIUM CANADENSE
Wood Sage
Coastal areas / July–August / 3 ft.

The upper lip of the flower of wood sage, instead of being hooded as in other mints, is split open and is not well developed. Thus, it appears as if the flower has only a lower lip with five lobes, above which project the four, equally long stamens and one pistil. The pink to purple flowers occur in rows at the ends of terminal branches. Stems and leaves are hairy and the leaves are lance-shaped and toothed.

THYMUS SERPYLLUM
Wild Thyme
Fields, roadsides, dry areas | June–September | Ground cover

This species of thyme is a small, hairy, creeping plant with erect branches and purple flowers. The flowers emerge in groups from the axils of paddle-shaped leaves. The sepals are fused for part of their length and they form two lips; they are also strongly veined and hairy. The petals are somewhat hairy on the outside of the tube as well as on the inside. There are four stamens of equal length and a long pistil.

TRICHOSTEMA DICHOTOMUM
Blue Curls
Woods, fields / August–September / 2 ft.

This is a sticky plant with blue flowers which are borne singly
at the ends of small branches. Each flower has a distinctive set
of four blue, curved stamens of equal length. The corolla tube
has five lobes: four upper lobes and one that bends sharply
downward. The sepal tube has three, sharply pointed upper
lobes and two shorter, lower lobes. The leaves are narrow,
without stalks, and smooth margined.

LAURACEAE *Laurel family*

This family consists of trees and shrubs with an aromatic taste or smell, and includes the spice bush, avocado, laurel, and cinnamon. The flowers are small and yellow and the fruit is a drupe.

SASSAFRAS ALBIDUM
Sassafras
Woods, roadsides / May / 40 ft.

Sassafras is generally a shrub or small tree with irregularly lobed leaves. The leaves are two- or three-lobed, but the pattern of lobes is so variable that it may be difficult to find two identical leaves on the same plant. There are separate male and female plants.

LENTIBULARIACEAE   *Bladderwort family*

Plants of this family are herbs which trap tiny animals that live in ponds and swampy areas. The flowers are small, irregular, and two-lipped; the lower lip extends back into a spur. There are two stamens and the fruit is a capsule.

UTRICULARIA INFLATA
Inflated Bladderwort
Ponds and ditches / July–August / 8 in.

The bladderworts are aquatic plants and they trap insect larvae and other small animals by means of bladders or sacs produced by the hundreds on finely dissected leaves and branches. The sacs lie below the surface of the shallow water, while the yellow flowers are on leafless stalks above the water. The leaf stalks of the inflated bladderwort are broad and help the plant to float. *Utricularia gibba* (humped bladderwort) has yellow flowers and it grows along the bottom of shallow ponds. It has very thin stems and only a few bladders. *Utricularia purpurea* (purple bladderwort) has pink-purple flowers. *Utricularia vulgaris* (common bladderwort) has yellow flowers and very leafy stems. It is free-floating, but its leaves are not inflated and the bladders are large and prominent.

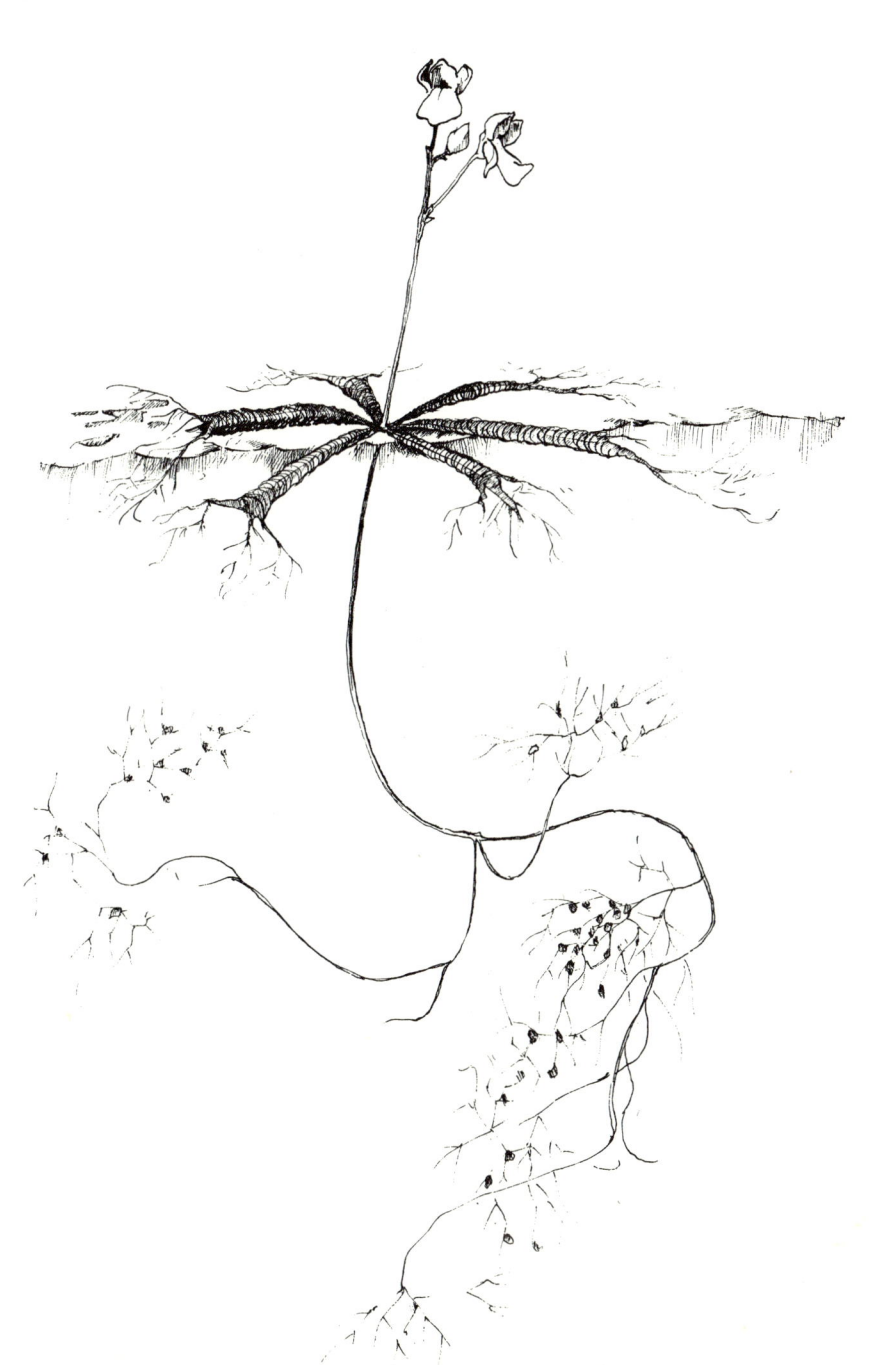

LINACEAE *Flax family*

The flax family of herbs has slender leaves and regular flowers whose parts are in sets of five. The fruit is a capsule.

LINUM VIRGINIANUM
Yellow Flax
Roadsides, woods | June–September | 3 ft.

This plant has a branched inflorescence with yellow flowers and oval leaves. The stem fiber of *Linum usitatissimum* (common flax) has been used since ancient times to make linen. The plant also has been used for the manufacture of linseed oil. The plant is found frequently growing in uncultivated fields and along roadsides. It has blue flowers, narrow, lance-shaped leaves and usually a single, unbranched stem.

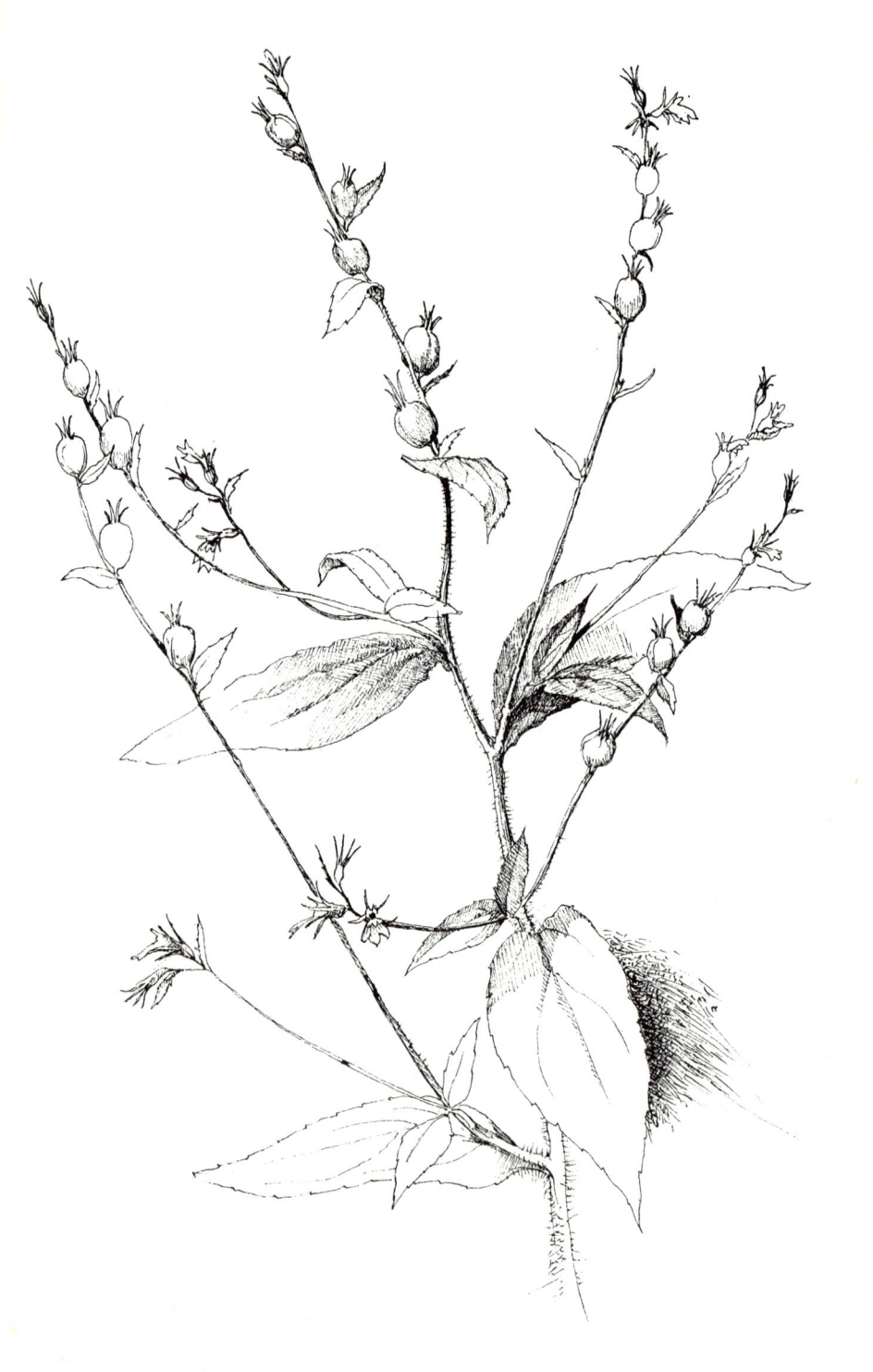

LOBELIACEAE   *Lobelia family*

Members of this family are herbs that are closely related to, or according to some botanists, actually part of the bluebell family (Campanulaceae). Their flowers are irregular with an upper and a lower lip. The flower parts are in fives and the fruit is a capsule. The plants have a milky, poisonous sap.

LOBELIA INFLATA
Indian Tobacco
Fields, roadsides / July–September / 3 ft.

This plant has characteristic inflated fruits which are found in the axils of the leaves along the ridged stem. The leaves are narrowly egg-shaped and slightly toothed and the stems and leaves are hairy. The lower lip of the flower has three-pointed lobes and a bearded patch on a white background near the mouth of the corolla tube. The upper lip of the flower has two narrow lobes that are more or less erect. The five anthers are fused around the style (the filaments are free), and the stamens and pistil arch forward in between the two lobes of the upper lip. The flowers are in racemes from the leaf axils. *Lobelia cardinalis* (cardinal flower) has bright scarlet flowers and grows along streams and in shallow rivers. *Lobelia kalmii* (brook lobelia) grows in wet areas and has widely spaced, blue flowers which have a white center, a basal set of spatulate leaves, and narrow, upper leaves. *Lobelia spicata* (pale-blue lobelia) grows in fields and has blue flowers on a terminal spike.

LYTHRACEAE   *Loosestrife family*

This family of herbs has axillary flowers with an expanded cup-shaped or tubular base. The number of sepals and petals varies between four and seven and the number of stamens is equal to or is twice as many as the number of petals. The fruit is a capsule.

DECODON VERTICILLATUS
Swamp Loosestrife
Swamps, ponds / July–August / 9 ft.

The stems of this plant generally are bent over and grooved. The leaves are lance-shaped and are usually in whorls of three to four. The undersides of the leaves are somewhat hairy, but the stems are smooth. Many purple flowers lie in the axils of the upper leaves and there are several types of flowers that differ in the length of the style. Each flower has five long and five short stamens, an urnlike floral base, five to seven short, spinelike sepals, and five long and narrow, separate petals.

LYTHRUM SALICARIA
Purple Loosestrife
Wet areas / July–August / 5 ft.

The numerous purple flowers of this loosestrife are arranged in long spikes at the top of the plant. The square stems and leaves are somewhat hairy. The leaves are opposite and lance-shaped and they clasp the stem; sometimes, there may be a whorl of three leaves. Each pair of leaves occurs at right angles to the pair above and below it. The individual plants may differ in the types of flowers they have; that is, some plants may have flowers with a long pistil and short stamens and other plants may have flowers with long stamens and a short pistil. The floral base is bell-shaped; there are four to six spinelike sepals, and four to six straplike and separate petals. Colorful masses of this plant grow in swampy depressions and in ponds along the roadsides. Found commonly with golden rods and joe-pye-weed, the combination of colors makes an excellent artistic subject.

MALVACEAE *Mallow family*

Members of this family have open flowers and alternate leaves. The flowers have five sepals and five petals which in the bud are rolled together lengthwise. There are numerous stamens whose filaments are fused to form a hollow column through which the styles protrude. The fruit is a capsule or a ring of separate carpels (five or more). Hollyhock, a common garden ornamental, and the cotton plant belong to this family.

ABUTILON THEOPHRASTI
Velvet-leaf
Fields, waste areas / August–October / 4 ft.

All parts of this plant are covered densely with soft, fine hairs. The leaves are broad and heart-shaped and usually taper to a pointed tip. The flowers are yellow and the fruit consists of a circle of ten to fifteen separate, hairy, mature carpels, each with a pointed tip, and three or more seeds.

HIBISCUS PALUSTRIS
Rose Mallow
Salt marshes, along streams in eastern
Massachusetts / July–September / 6 ft.

The pink flowers of rose mallow are large and showy and have a prominent style with five branches, each of which has a round, hairy stigma. Many yellow stamens project from the sides of the hollow, central column. The leaves are large, egg-shaped, and hairy underneath and the upper part of the stem is also hairy. The fruit is a capsule.

MALVA NEGLECTA
Common Mallow
Waste areas, around buildings / June–September

The common mallow is a weedy, creeping plant. It has round leaves on long stalks, and white to purplish flowers. *Malva moschata* (musk-mallow) grows up to three feet in fields and along roadsides, and it has leaves with five or more narrow segments. The flowers are white to light purple. They arise from the leaf axils along the stem and also occur in groups at the top of the stem.

MELASTOMACEAE   *Melastome family*

One species of this primarily tropical family grows commonly in Massachusetts.

RHEXIA VIRGINICA
Meadow Beauty
Sandy fields, along ponds / July–September / 18 in.

The purple flowers of meadow beauty have eight stamens with large, yellow anthers which are bowed and attached to their filaments at an angle. There are four petals and four sepals and the base of the flower (hypanthium) is vaselike and hairy. The leaves are opposite, bristly, and without stalks, and four prominent ridges run the length of the stem. The roots contain small tuberlike thickenings.

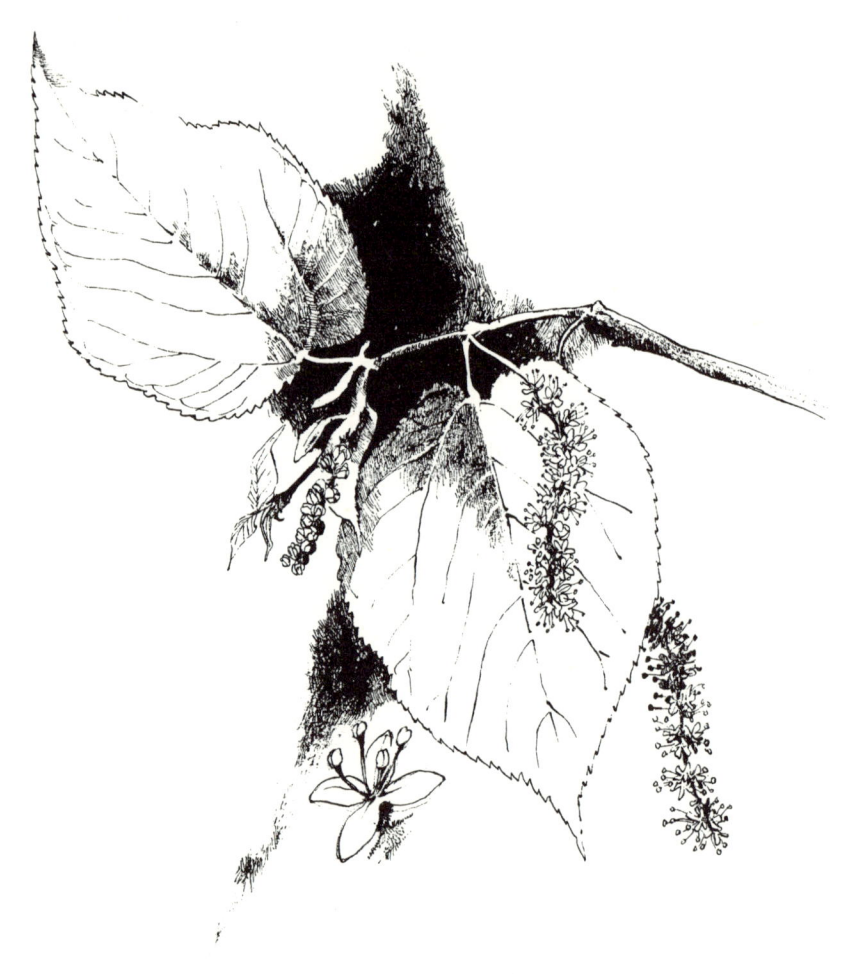

MORACEAE   *Mulberry family*

This large and important family of mostly tropical trees includes species which produce the fig and the breadfruit. The plants have a milky sap and small, unisexual flowers with four sepals but no petals.

MORUS RUBRA
Red Mulberry
Calcareous soil in woods / April–June / 60 ft.

The red mulberry is a tree with prominently toothed leaves of different shapes. Some leaves are broad and egg-shaped while others are cut deeply into several lobes. The underside of the leaves is hairy and the fruit is reddish-black. *Morus alba* (white mulberry) is similar in appearance but it has hairless leaves and white to pinkish fruits. The structure called the fruit of the mulberry is actually a multiple fruit that resembles a blackberry; i.e., it is a tight cluster of small, one-seeded fruits (achenes), each of which is surrounded by a fleshy calyx. The edible fruits of the red mulberry are prized by birds and humans. The leaves of the white mulberry have long been used as a source of food for silkworms.

MYRICACEAE   *Bayberry family*

This family of shrubs has small, unisexual flowers which lack
sepals and petals and are crowded into small, round or elongated catkins. The leaves are aromatic when crushed and they are
sprinkled with small, resinous droplets.

MYRICA ASPLENIFOLIA
Sweet Fern
Dry, open areas, fields, along roadsides / April–May / 3 ft.

Sweet fern is a low-growing, hairy shrub with fernlike leaves.
The male and female flowers may occur on the same plant
or on separate plants and appear before the leaves. The staminate catkin is long and pendulous while the pistillate catkin is
small and round. The fruit resembles a bur and consists of
small nuts surrounded by the bracts that were around the pistillate flowers.

MYRICA GALE
Sweet Gale
Bogs, swamps, along ponds / April–May / 3 ft.

Sweet gale is a low-growing shrub with wedge-shaped leaves
and erect, brown branches. The male and female catkins are o
separate plants and appear before the leaves. The male catkin
is cigar-shaped when it is young; when it matures, its scales
open wide and reveal the stamens attached to the central axis
The female catkin is small and round; when mature, its styles
hang out from between the scales like red threads. The fruits
are conelike.

MYRICA PENSYLVANICA
Bayberry
Along the seashore, and as far inland as
Worcester County / May–June / 6 ft.

Bayberry is a thick shrub with whitish-gray bark and fruits
(nuts) coated with a gray-white wax. The flowers and fruits
occur along the stem below the upper leafy branches. Dried
branches with their numerous grapelike clusters of waxy fruits
are especially suited for indoor decoration.

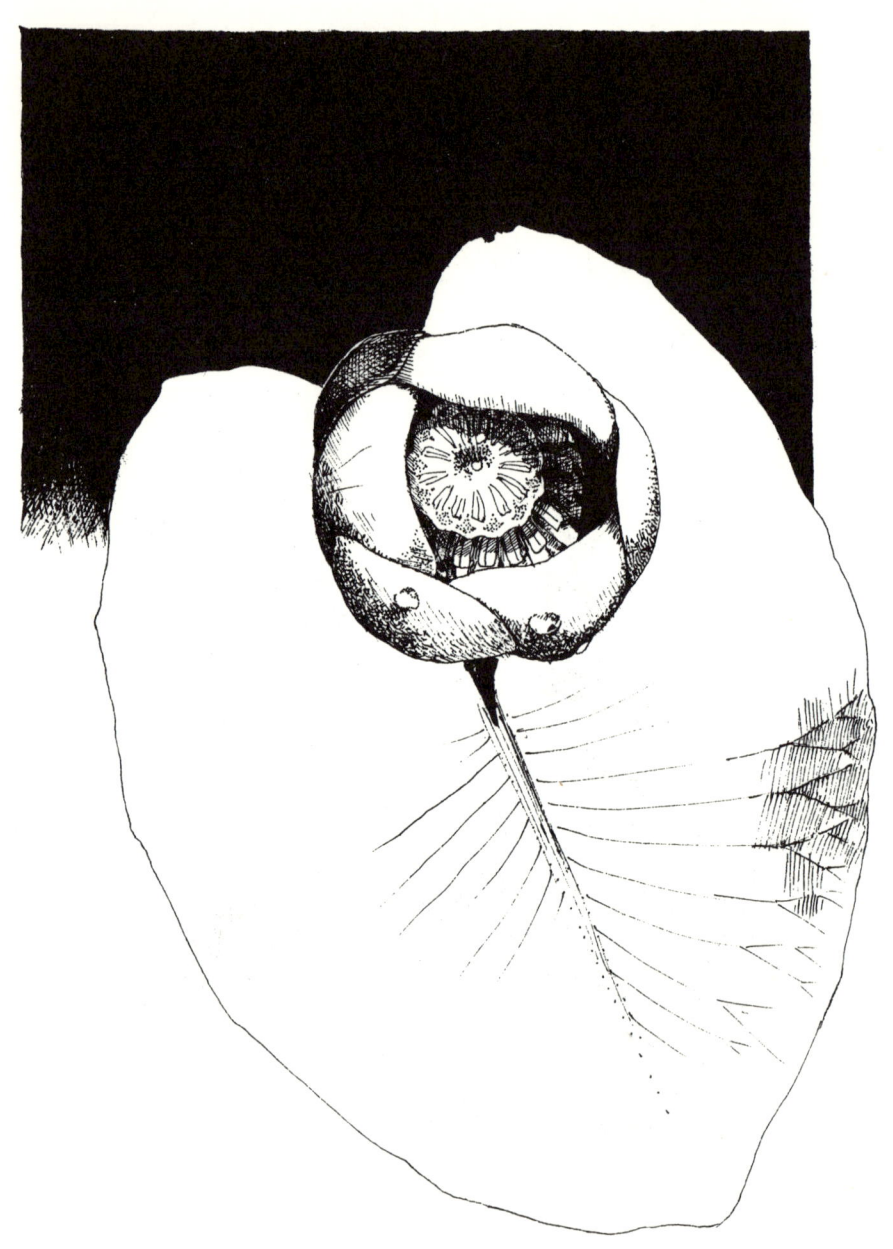

NYMPHAEACEAE   *Water Lily family*

This family of aquatic herbs has tuberous stems buried in the mud of shallow ponds and lakes. The leaves and flowers are situated on long stalks either on the water's surface or above it.

NUPHAR VARIEGATUM
Yellow Pond Lily
Ponds | May–August

The heart-shaped leaves of this familiar pond lily float on the surface of the water. The flowers are on separate stalks and each flower consists of five to six yellow, petallike sepals which are red on their inner lower parts. The petals are small and numerous and are located along the base of the sepals. There are also many stamens which are taller than the petals. The pistil is large with a thick style and a broad, round stigma with many rays on its surface. *Nymphaea odorata* (white water lily) has round leaves which float on the surface of the water. The leaves are purple on their underside and they have a v-shaped notch at the point where the stalk attaches to the blade. The white-pink flowers are fragrant and have four green sepals and numerous petals which decrease in size from the outer to the inner part of the flower. The flowers open in the morning and close at night. There are many stamens, the outer ones being petallike, and the ovary is large and thick with many stigmatic rays that curve inward.

OLEACEAE *Olive family*

This family of trees and shrubs includes such familiar ornamentals as forsythia and lilac as well as the commercial species
of olive. The small flowers occur in clusters and the opposite
leaves are either simple or compound.

FRAXINUS AMERICANA
White Ash
Woods / April–June / 80 ft.

White ash is an important timber tree. Each of its compound
leaves has five to nine stalked leaflets, and the undersurface
of the leaves is whitish. Flowers and fruits are clustered along
separate stalks which emerge from the axils of the previous
year's crop of leaves. Male and female flowers occur on separate trees; the flowers appear before or at the same time as the
leaves. The fruit is a samara with a long and narrow wing.
Fraxinus nigra (black ash) occurs in swamps and has leaflets
without stalks.

ONAGRACEAE   *Evening Primrose family*

Members of this family are herbs with simple opposite or alternate leaves, and most of the species have their floral parts in sets of four. The number of stamens equals or is twice as many as the number of petals. The hypanthium generally forms a long tube at the base of which is the ovary. The fruit is a capsule.

CIRCAEA QUADRISULCATA
Enchanter's Nightshade
Woods / June–July / 3 ft.

This is an unusual plant in two respects. First, the distinctive fruits are covered with bristles and resemble little burs; they are more noticeable than the small, white flowers that are arranged in a raceme along the stem. Second, the number of flower parts is fewer than most plants, i.e., two sepals, two deeply lobed petals, and two stamens. The leaves are opposite and have long stalks. *Circaea alpina* (small enchanter's nightshade) grows up to one foot in swamps and wet woods. It has few flowers and heart-shaped leaves with large, marginal teeth.

EPILOBIUM ANGUSTIFOLIUM
Fireweed
Clearings in woody areas, burned-over
areas | June–August | 6 ft.

Firweed is an erect, stately plant with crowded, wrinkled
leaves which are lance-shaped and alternate. A long flower
stalk extends beyond the leaves. On the lower part of the flower stalk are long and narrow, upward-turning seed pods; above
the pods are open flowers and above them, to the tip of the
stem, are drooping flower buds. The stem, flowers, and fruits
are reddish-purple. Each flower has four narrow sepals which
alternate with four broader and unfused petals, a long ovary,
eight stamens, and a hanging pistil. The sepals are darker than
the petals. These plants grow together in large, colorful groups.
The fruits open by means of four segments which curl backward and reveal many seeds, each of which has a small parachute of white hairs. *E. angustifolium* is distinguished from the
other common species of this genus by its four-lobed stigma.
The other species have an undivided stigma. *Epilobium colora-
tum* (red-leaved willow-herb) grows up to three feet in bogs
and wet woods, and it is hairy and bushy; it has many pink to
white, small flowers, long, narrow fruits, and leaves with marginal teeth. *Epilobium leptophyllum* (narrow-leaved willow-
herb) also grows in bogs, grows up to four feet, and has hairy,
narrow leaves and pink to white flowers. Its leaves do not have
marginal teeth. *Epilobium strictum* (downy willow-herb)
grows in bogs, reaches up to two feet, and has a hairy, unbranched stem with narrow, smooth-margined leaves and small,
pink flowers.

OENOTHERA BIENNIS
Common Evening Primrose
Fields, roadsides / July–September / 6 ft.

The common evening primrose is a leafy, biennial plant with yellow flowers on the upper part of the plant. The stem is thick, tinged with red, and has white hairs. The bracts at the base of the flowers resemble the lance-shaped leaves but they are smaller. The flowers have four reflexed sepals, four large petals, and eight stamens of equal length. The stigma has four large, fingerlike lobes. *Oenothera parviflora* (small-flowered evening primrose) grows up to two and one-half feet and it has small, yellow flowers. *Oenothera perennis* (sundrops) is a small, slender plant with square, winged ovaries and capsules. The other two species have round ovaries and capsules. The flowers of *O. perennis* open in the sunlight while the flowers of the other two species open only at night.

OROBANCHACEAE   *Broom-rape family*

This interesting family consists of herbs whose members have
lost their green color and have become parasitic on the roots
of other plants. The leaves are scalelike and the flowers have a
tubular, two-lipped corolla. There are four stamens—two short
and two long—and the fruit is a capsule.

CONOPHOLIS AMERICANA
Squaw-root
Parasitic on roots of oak trees | June–July | 6 in.

The stem of squaw-root is surrounded by many overlapping,
fleshy, yellow-brown scales. The flowers are also yellow-brown
and arise from the axils of the scales along the upper half of
the plant.

EPIPHEGUS VIRGINIANA
Beech-drops
Parasitic on the roots of beech trees / August–October / 12 in.

Beech-drops is a many-branched plant with light brown stems, a bulbous base, and two kinds of flowers. The flowers on the upper part of the plant are white with purplish-brown stripes and they are sterile. Flowers on the lower part of the plant are fertile, but they never open. The corolla forms a pointed cap on top of the ovary as it enlarges on the lower flowers.

OROBANCHE UNIFLORA
Cancer-root
Woods, parasitic on the roots of
different plants / May–July / 10 in.

The stem of this pale plant is short and is mostly buried in the ground. Each hairy flower stalk has one flower which is white to lilac with yellow streaks inside the tube. There are two yellow ridges on the lower lip of the flower, five sharply pointed sepals, and five petals fused into a two-lipped tube.

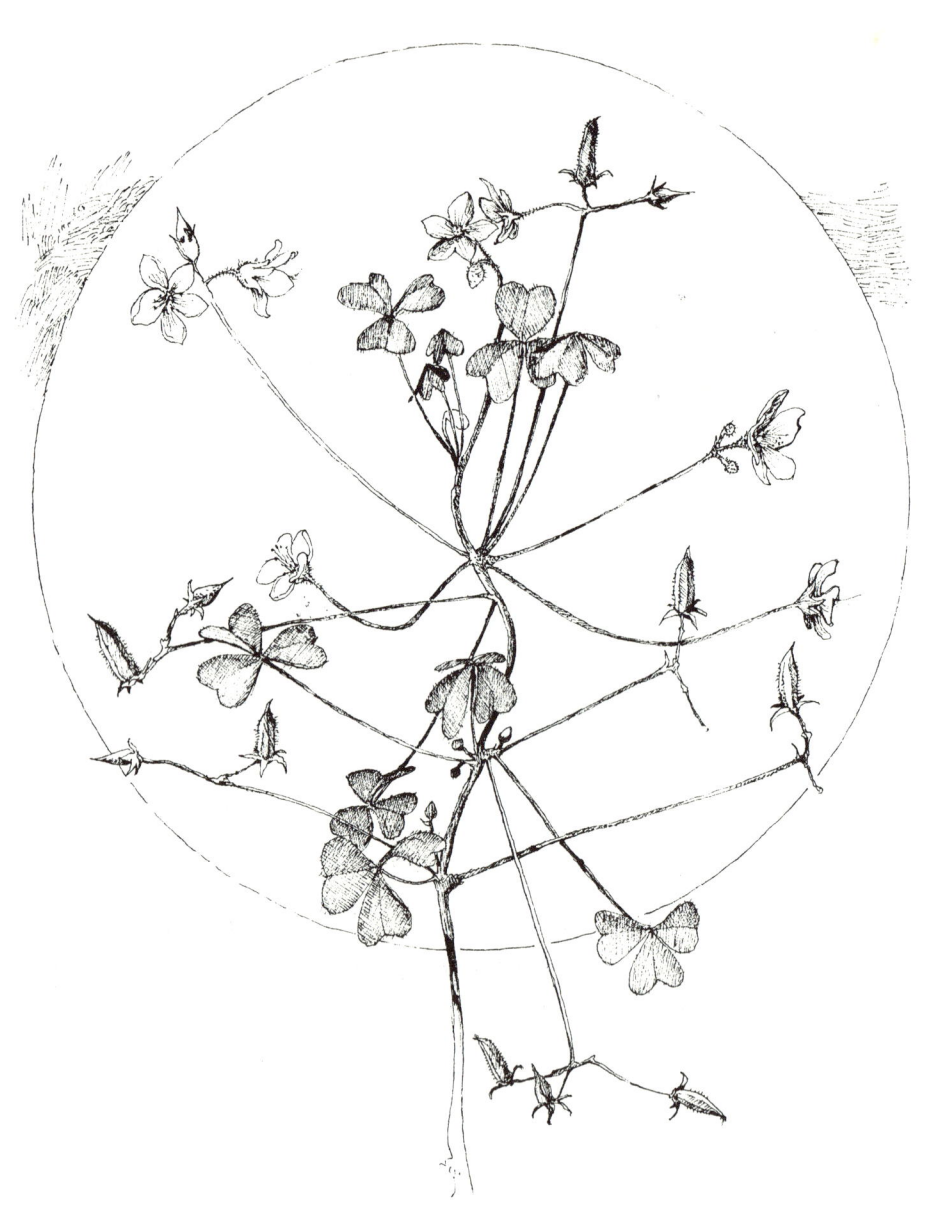

OXALIDACEAE   *Wood Sorrel family*

The species of this family are small herbs whose flowers have
five sepals, five petals, ten stamens, and one pistil. The leaves
are subdivided into three heart-shaped leaflets which have a
characteristic folded appearance at night. The fruit is a capsule.

OXALIS EUROPAEA
Common Wood Sorrel
Gardens, lawns, roadsides / June–August / 3 ft.

The common wood sorrel is an erect or sprawling, hairy plant
which may have many branches. Its flowers are yellow and the
fruit stalks are not bent downward. *Oxalis stricta* (yellow
wood sorrel) is similar, but it grows in fields and its fruit stalks
bend sharply downward, although they still bear fruits in an
upright position. *Oxalis montana* (mountain wood sorrel) is a
creeping plant found in woods at higher elevations; it has
white flowers with pink veins.

PAPAVERACEAE   *Poppy family*

The poppy family consists of herbs with a colored or milky sap. The flower has two sepals (which fall off soon after the bud opens), four to twelve petals, many stamens, and one pistil. The fruit is a capsule. Species of *Papaver* include the opium poppy and the oriental poppy, which are rarely found outside of cultivated areas.

CHELIDONIUM MAJUS
Celandine
Roadsides, waste areas / May–June / 2 ft.

The celandine has a yellow sap and bright yellow flowers. The flower has four petals, and stamens whose anthers are attached to the filaments by threadlike connections. The pistil has a long, flat ovary and a small style. The leaves are pinnately compound and the stems are hairy. The fruit is an elongated capsule.

SANGUINARIA CANADENSIS
Bloodroot
Woods / April–May / 12 in.

The flower of bloodroot typically has eight snowy-white petals which close in the evening and open widely in full sunlight. The flower and leaf arise on separate stems from an underground fleshy rhizome. Each stem bears either one flower or one leaf. The leaf is pale green, lobed, and has prominent veins. As the plant emerges from the ground, the leaves are wrapped protectively around the flower bud. When the bud opens, the leaf unfolds; after flowering is over, the leaf often becomes very broad. The blood-red sap of the plant is especially concentrated in the rhizome. The petals and yellow stamens fall off after a few days and only the single pistil remains at the end of the stem. These plants form large communities in some woodland areas.

PHYTOLACCACEAE   *Pokeweed family*

This primarily tropical family has only one representative in Massachusetts.

PHYTOLACCA AMERICANA
Pokeweed
Roadsides, waste areas | July–September | 9 ft.

Pokeweed is a coarse, weedy, poisonous plant with widely spread branches and stout, reddish stems. The small, greenish-white flowers are arranged in long racemes which emerge opposite the leaves. There are no petals, but there are five petallike sepals, generally ten stamens, and ten, fused pistils. The fruit is a dark purple berry.

PLANTAGINACEAE   *Plantain family*

The plantains are weedy plants with basal leaves and small flowers. Representatives of this family are persistent pests of lawns and cultivated land.

PLANTAGO RUGELII
Pale Plantain
Roadsides, lawns / June–September / 18 in.

The pale plantain has long flower stalks and a basal rosette of leaves. The leaves are broad and oval, or elliptic with a smooth, trough-shaped, purple petiole. The tiny green flowers are bisexual and are spaced along long spikes. *Plantago aristata* (bracted plantain) has narrow, linear leaves with long, hairy bracts around the flowers. *Plantago lanceolata* (English plantain) has clusters of long, narrow, lance-shaped leaves and long stalks at the end of which are short spikes. The tiny flowers are tightly grouped on the spikes. Flowering proceeds from the bottom of the spikes upward and only a few flowers are in bloom at the same time. *Plantago major* (common plantain) is similar to *P. rugelii* except that its petioles are hairy along their lower sides, and they are green and not tinged with purple.

PLATANACEAE   *Plane Tree family*

This family of trees has tiny flowers which are grouped tightly in round, ball-shaped heads. The male and female flowers occur on separate heads. The fruit is an achene surrounded by long hairs and many achenes develop from the pistils on the female head. The London plane tree is a commonly planted hybrid in the eastern United States.

PLATANUS OCCIDENTALIS
Sycamore
Along rivers / May–June / 130 ft.

The sycamore is one of this state's largest trees. It has a distinctive mottled, greenish-white trunk. Its leaves are large and broad with sharply pointed lobes. The enlarged base of the petiole encloses next year's bud.

POLYGALACEAE   *Milkwort family*

This small family of herbs has highly irregular flowers and simple, alternate or whorled leaves.

POLYGALA PAUCIFOLIA
Fringed Polygala
Wet woods / April / 6 in.

The flowers of fringed polygala are rose purple, sometimes white, and resemble miniature orchids. Each flower has five sepals: three are small, while two are large and petallike and form prominent lateral wings. The three petals join into a tube and the lower petal forms a keel-shaped pouch within which are six stamens and one pistil. The tip of the lower petal is subdivided into small corallike segments. This plant also produces insignificant basal flowers which do not open but which are fertile. The plants arise from slender, purplish rhizomes which grow just below the leafy cover of the forest floor. The basal leaves on the stem are small, while those near the top of the plant are larger and clustered in groups of three to six. The leaves persist through the winter.

POLYGONACEAE   *Smartweed family*

This family of weedy herbs has simple leaves, small flowers, and stipules which surround the generally swollen nodes of the stem. The fruit is an achene. The smartweed family includes cultivated buckwheat, rhubarb, and species of sorrel or dock. Sheep sorrel (*Rumex acetosella*) is a very common weed of lawns and fields. It has arrow-shaped, green or reddish-brown leaves. Curled dock (*Rumex crispus*) is a stout plant with a thick stem and curly leaf margins, and is common along roadsides and in fields. There are about forty species of *Polygonum*, a summer- and fall-blooming genus, and many of the species are separated on the basis of small technical traits.

POLYGONUM CUSPIDATUM
Japanese Knotweed
Roadsides / August–September / 9 ft.

The Japanese knotweed is an extremely difficult weed to eradicate because of its vigorously growing network of underground rhizomes. Mature plants are shrubby and have groups of thick stems which resemble a miniature bamboo forest. The greenish-white flowers are borne on branched and erect racemes along the upper parts of the separate male and female plants. The leaves are broad and egg-shaped with a pointed tip.

POLYGONUM PENSYLVANICUM
Pinkweed
Fields, waste areas / July–September / 4 ft.

The pinkweed is erect and upward-growing and has branched stems with glandular hairs on their upper parts. Its pink flowers occur in many small, crowded, and erect racemes. The leaves are lance-shaped and they taper to a long point. They form a sheath around the stem. The achenes are small, black, and lens-shaped.

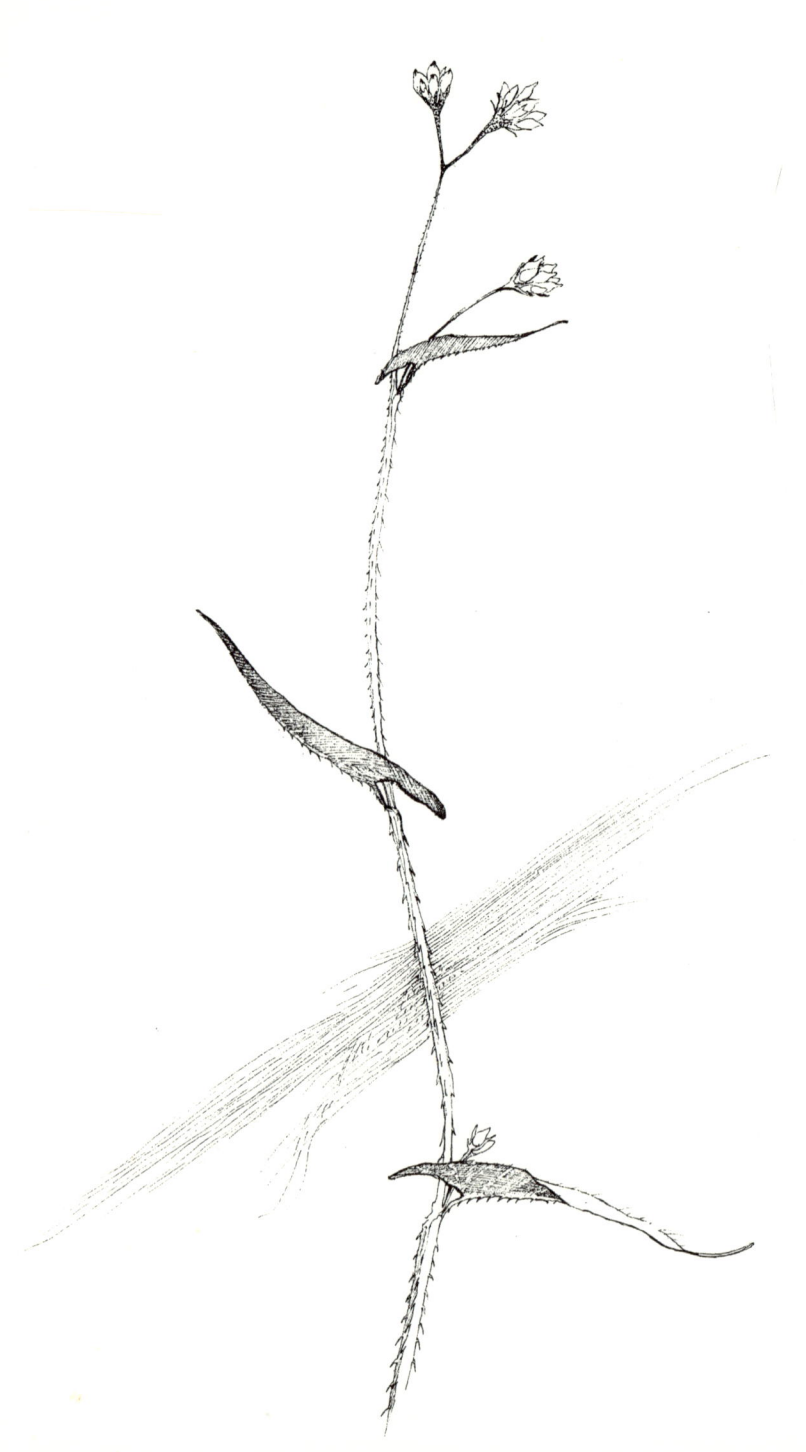

POLYGONUM SAGITTATUM
Arrow-Vine
Meadows, marshes / July–September / 6 ft.

This slender plant grows like a vine over other vegetation. Its stems and petioles are spiny and its leaves are generally arrow-shaped and widely spaced along the stems. The leaves also may be elliptical, and the achenes are small, dark, and three-sided. The pink to white flowers occur in small heads which are closely grouped together in small racemes. The inflorescences occur at the ends of long stalks.

TOVARA VIRGINIANA
Jumpseed
Woods / August–September / 4 ft.

The greenish white flowers of jumpseed are found in long, terminal racemes which project well beyond the lance-shaped
leaves. The pair of stipules at the base of the leaves tend to
overlap at their tips. The achenes are lens-shaped and they
retain two hooked styles on their upper ends.

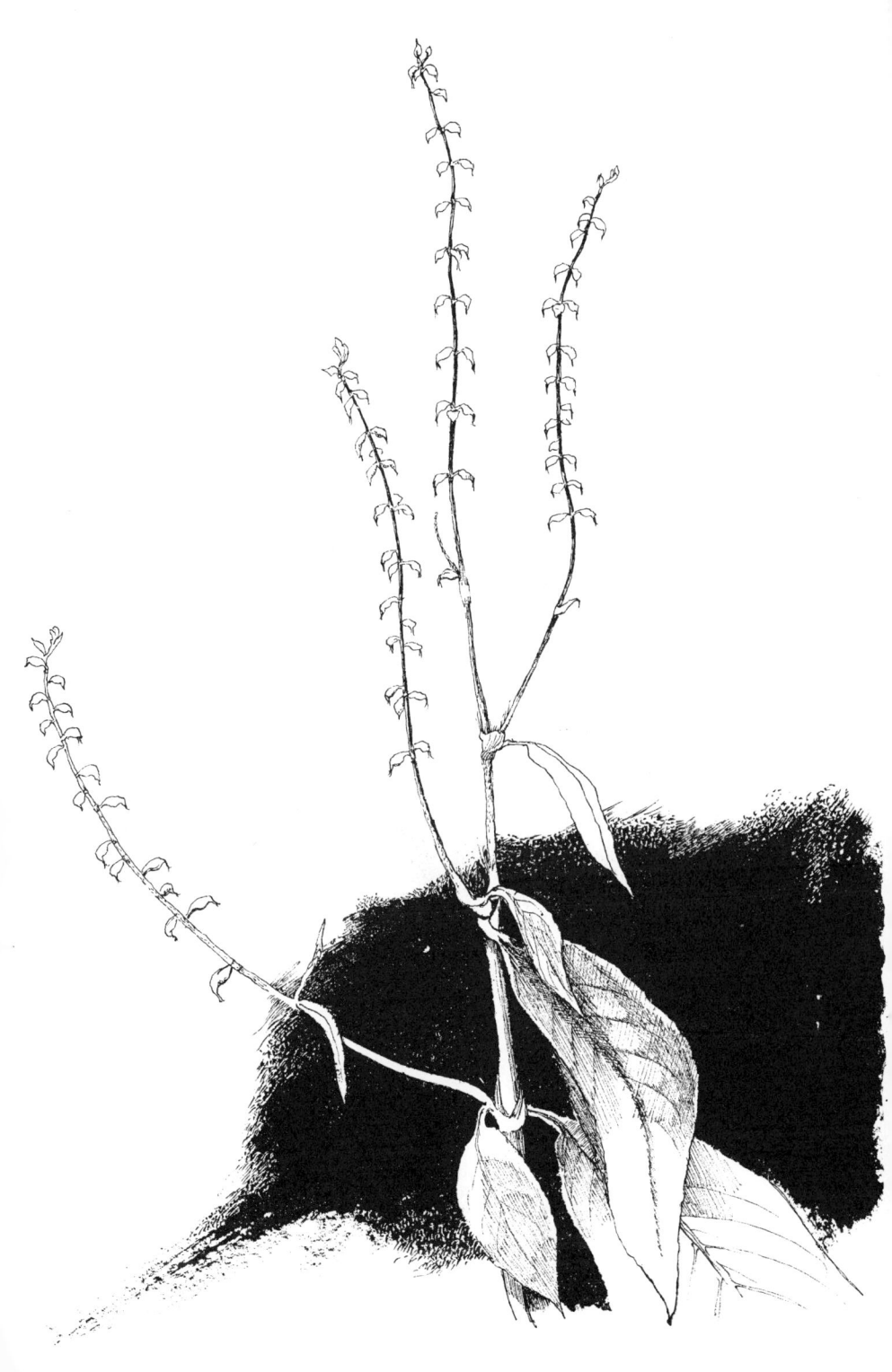

PORTULACACEAE   *Purslane family*

This family of herbs has flowers with two sepals, three to five petals, and a variable number of stamens. The fruit is a capsule.

CLAYTONIA CAROLINIANA
Spring Beauty
Woods / May–June / 12 in.

This plant originates from a deep, underground tuber from which the stem easily breaks off when pulled. Each stem has a pair of opposite leaves and a cluster of fragrant, white to pink flowers. Fruits develop on the lower part of the flower stalks while the upper flowers are still in bloom. Each flower has two brown sepals which remain around the fruit as it develops, five candy-striped petals, five stamens—one opposite each petal— and a long pistil with a three-lobed style.

PRIMULACEAE   *Primrose family*

This family of herbs is found commonly in the northern regions of the earth. The flower parts generally are in sets of five, the petals are fused to some degree, and the stamens are opposite the petals. The leaves are simple and the fruit is a capsule.

ANAGALLIS ARVENSIS
Pimpernel
Roadsides, fields / June–August / 12 in.

The pimpernel is a fair-weather plant because its flowers close on cloudy days and open on sunny days. The plant is usually branched and grows either upright or along the ground. The flowers are scarlet, or sometimes white or blue. The leaves lie in opposite pairs along the stem and the flowers occur on the ends of long stalks.

LYSIMACHIA CILIATA
Fringed Loosestrife
Swamps and along banks of streams / July–August / 3 ft.

The yellow flowers of this plant bend downward and their petals have sharp, marginal teeth. The petioles are densely hairy and the leaves are opposite and egg-shaped. Several small leaves occur at the axils of the larger leaves. The stems are often branched at their upper ends. The nodding flowers on long stalks make this plant easy to recognize.

LYSIMACHIA TERRESTRIS
Swamp Candles
Swamps, wet areas / June–August / 2 ft.

This plant has one long, terminal raceme of flowers. The flowers are yellow with dark streaks or dots and together with their associated bracts they crowd the raceme. The leaves are opposite and lance-shaped. *Lysimachia quadrifolia* (whorled loosestrife) is common in fields and open woods. It has characteristic tiers of whorled leaves from the axils of which emerge a circle of four to five long, stalked flowers. The petals and sepals are streaked with short red stripes and the center of the flower, around the ovary, has a dark red ring.

TRIENTALIS BOREALIS
Star Flower
Woods / May–June / 8 in.

The stem of the star flower arises from a slender tuberous rhizome. At the upper end of the stem is a whorl of leaves from which arise one or several individual flowers, each on a slim stalk. The lower parts of the stem contain a few isolated scaly leaves. The white flowers are open and star-shaped and have five to nine sharply pointed petals and a similar number of thin, green sepals. The stamens are joined at their base to form a short ring around the ovary of the single pistil.

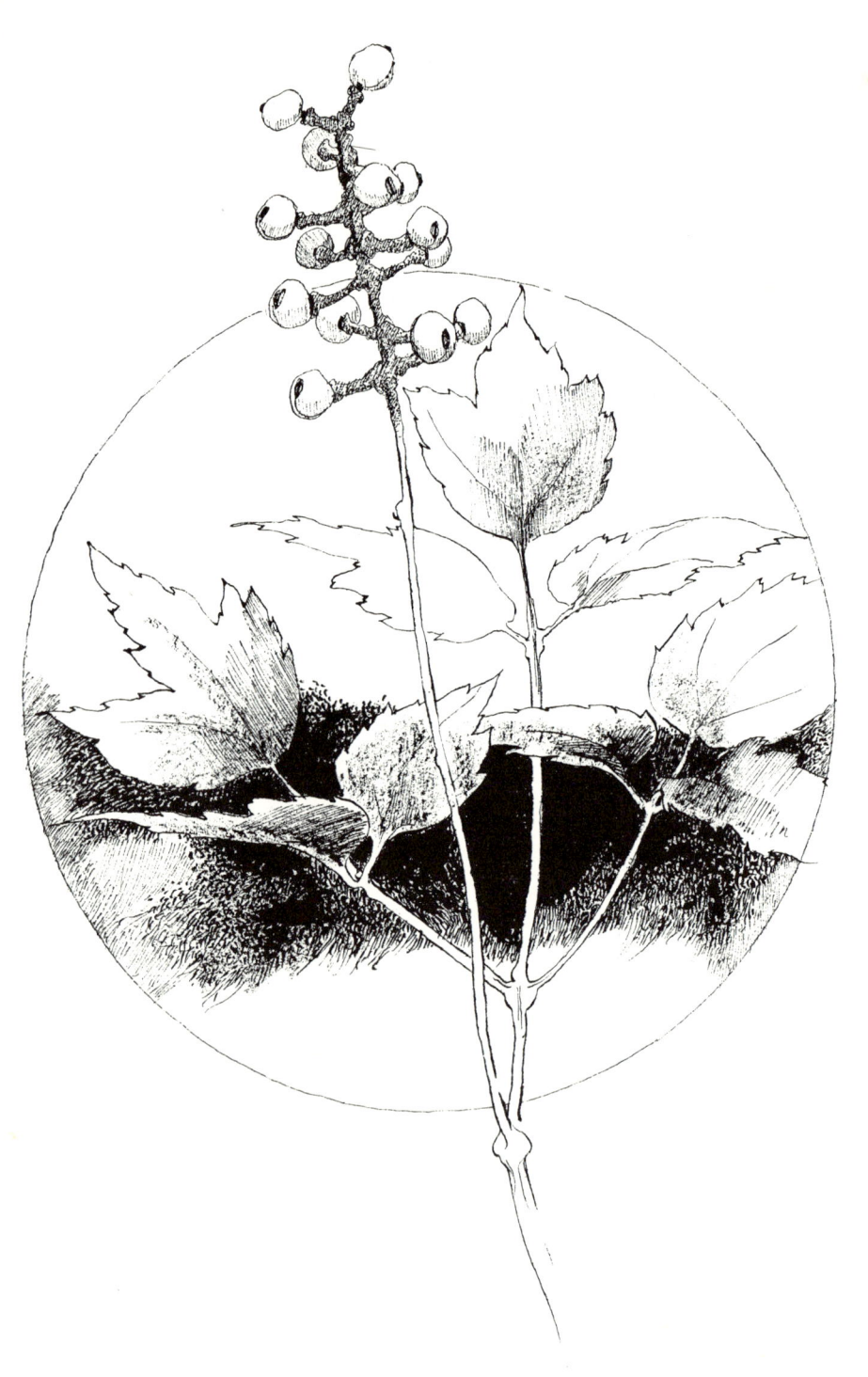

RANUNCULACEAE   *Buttercup family*

Members of this large family of herbs are a distinctive part
of our spring and early summer flora. The flowers are characterized by many stamens and one to many simple pistils, and
the flower parts are free from each other. Sepals are generally
present, but those species that lack petals have sepals that are
petallike. Botanists can distinguish between sepals and petals
not only from their position on the flower but also from their
number of veins. Sepals have three or more veins while petals
generally have only one. The fruit is either a one-seeded
achene, a berry, or a follicle.

ACTAEA ALBA
White Baneberry
Woods / May–June / 2 ft.

The leaves of white baneberry are large and usually subdivided
several times into sets of three; the margins of the leaflets are
sharply toothed. The small flowers occur in a raceme at the
end of a stalk. Each flower has a single pistil, but the other
floral parts vary in number. The wide stigma of the pistil sits
over the ovary like a cap and persists on the berry as a red-purple spot. Numerous stamens with long filaments radiate from
the flower. The petals are flat and narrow and taper toward
their base. The fruits are white berries with red stalks. *Actaea
rubra* (red baneberry) has red berries; its stigma is not as wide
as the ovary.

ANEMONE QUINQUEFOLIA
Wood Anemone
Woods / April–June / 10 in.

The flower of the wood anemone is on a single stalk which arises from a whorl of three leaves. Each leaf has three leaflets, but the two lateral ones are cut so deeply that the plant appears to have five leaflets. There are no petals, only five white, sometimes pink-tinged sepals. The flower has many white stamens and green pistils which mature into achenes.

ANEMONE VIRGINIANA
Thimbleweed
Woods / July–August / 2 ft.

This tall anemone has hairy leaves and stems. The leaves are in several groups of two or three and each group forms a whorl around the stem. The leaves are subdivided into three main lobes with toothed margins, and some of the lobes may be partially subdivided. Each flower lies at the end of a long stalk which arises from a cluster of leaves. There are no petals, but the sepals are white, petallike, and are generally five in number. The tip of the flower stalk (receptacle) is shaped like a thimble; to it are attached the many pistils and, later, the achenes. The mature achenes are surrounded by a woolly coating.

ANEMONELLA THALICTROIDES
Rue Anemone
Woods / April–May / 10 in.

The rue anemone has several small, underground tubers. Its leaves are subdivided into three leaflets which have rounded lobes. The white to pink flowers are borne singly on long stalks, and they have only petallike sepals. After fertilization, the sepals fall off and leave behind a group of achenes.

AQUILEGIA CANADENSIS
Wild Columbine
Rocky fields, woods / May–June / 3 ft.

The hanging red flowers of wild columbine have a yellow center, five red sepals, and five tubular petals which taper to small
nectar-containing bulbs. Numerous, long yellow stamens and
five equally long and slim pistils project beyond the flower.
Some of the stamens near the center of the flower are sterile;
i.e., they are without anthers and their filaments are flat, mem-
branelike, and wider than those of the normal fertile stamens.
The leaves are divided into three leaflets, each of which in the
lower leaves may be subdivided into threes or have three lobes.
The fruit is a follicle.

CALTHA PALUSTRIS
Cowslip
Wet meadows, along streams / April–May / 2 ft.

The familiar yellow flowers of cowslip brighten many wet areas of our state. The stems are hollow and the leaves are large and round. Leaves near the base of the plant have long petioles, while those along the upper parts of the plant have practically no petioles. The flowers have five to nine petallike sepals, many stamens, and pistils which mature into clusters of follicles.

435

COPTIS GROENLANDICA
Gold Thread
Woods, swamps / May–June / 6 in.

This small, slender plant has single flowers at the ends of long, leafless stalks. Each flower has four to seven white petallike sepals tinged with purple on their outer surfaces, and numerous white stamens. Among the stamens are five to seven lollipop-shaped structures which have a white stalk and a hollow, yellow, nectar-producing head. Three to seven pistils form a circle around the center of the flower; each pistil is elevated on a stalk and has a curved style. The evergreen leaves occur on long petioles, and every year new leaves develop from a slim, orange rhizome. Each leaf is subdivided into three leaflets. The fruit is a follicle.

HEPATICA AMERICANA
Hepatica
Open, dry woods | April | 6 in.

The leaves of hepatica are evergreen, with three rounded lobes, and the new leaves and flower stalks are hairy. The flowers are purple, blue or white, sometimes pink, and they appear before the new leaves. There are five petallike sepals and many stamens and pistils. Three bracts occur at the base of each flower and they usually persist during development of the achenes.

RANUNCULUS ACRIS
Buttercup
Meadows, fields / May–August / 3 ft.

Our common buttercup is a tall, hairy plant with shiny, bright yellow petals. Leaves on lower parts of the stem are more abundant and better developed; they are cut deeply into three to seven main segments which in turn are divided into lobes. *Ranunculus abortivus* (small-flowered crowfoot) has small, inconspicuous flowers whose sepals are longer than the petals. The leaves at the base of the plant are round with scalloped margins while those higher on the stem are cut into narrow lobes. *R. bulbosus* (bulbous buttercup) is a shorter plant than *R. acris*, about one foot high, with a bulblike base of stems just below the ground. The plant is also hairy and the sepals of the flower are bent backward. The fruits of these buttercups are achenes.

THALICTRUM POLYGAMUM
Tall Meadow-rue
Wet meadows / July–August / 5 ft.

The leaves of tall meadow-rue are subdivided several times, each time into threes, with each of the final three leaflets having three lobes. Flowers are white and occur in loosely branched groups at the top of the stem. There are no petals, and the petallike sepals fall off early, leaving flowers which have only numerous, upright stamens with white filaments and yellow anthers. Some flowers have a pistil as well as stamens. *T. dioicum* (early meadow-rue) blooms earlier (May–June), is shorter (1 to 2 feet), and has hanging flowers with purplish sepals. The fruits of both plants are achenes.

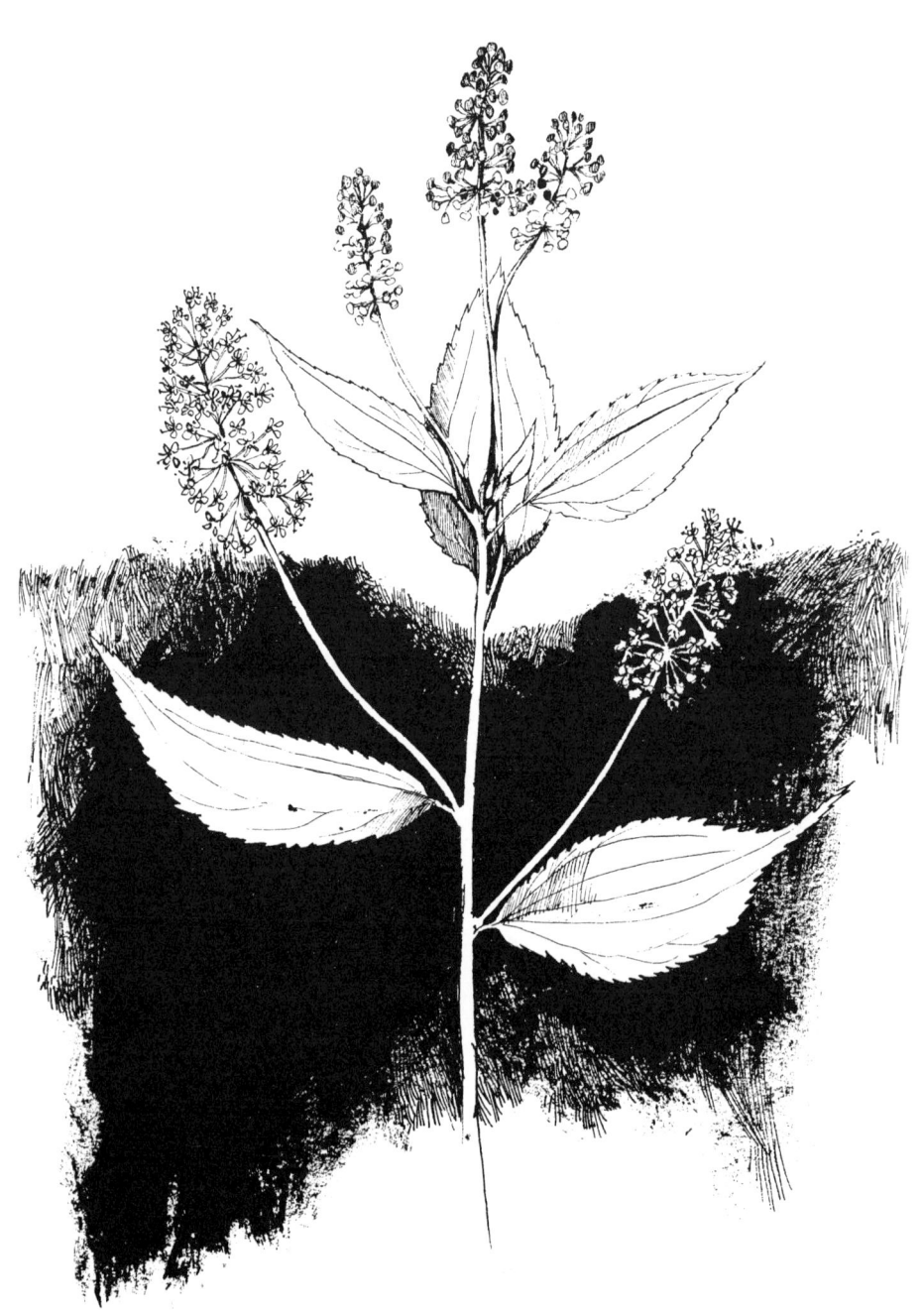

RHAMNACEAE   *Buckthorn family*

A family of shrubs and small trees with simple, alternate leaves, and small flowers.

CEANOTHUS AMERICANUS
New Jersey Tea
Woods / July–August / 6 ft.

New Jersey tea is a small, branching shrub with sharply pointed, oval leaves which in earlier times were used to make tea. The white flowers are grouped on separate stalks. The fruit is a dry, blackish drupe and when it opens, three, small white nuts are revealed.

RHAMNUS FRANGULA
Buckthorn
Woods, roadsides / May–July / 21 ft.

This large shrub has escaped from cultivation and has spread
rapidly through the northeastern United States. The leaves are
smooth margined, and greenish flowers occur in small groups
in the leaf axils. The fruit is a drupe which is red at first and
then turns black. Each fruit has two to three seeds, and is re-
puted to have cathartic properties. Flowers and fruits may be
present on a shrub at the same time.

ROSACEAE   *Rose family*

This is a large family of herbs, shrubs, and trees. Many species are cultivated as ornamentals and for their edible fruits. The flowers have a flat, open appearance; they have five sepals, five unfused petals, many stamens, and one to many pistils. The end of the flower stalk (receptacle) is expanded into a saucer- or cup-shape, or it is enlarged into a mound shape. This expanded structure is called a hypanthium and to it are attached the parts of the flower. The leaves are compound or simple and they often have stipules. A variety of fruit types is produced, i.e., pome, achene, drupe, and follicle.

ARONIA PRUNIFOLIA
Chokeberry
Swamps / May–June / 10 ft.

The chokeberry is a shrub with oval leaves. The simple leaves have a hairy underside and many small teeth along their margins; along the midrib of the leaf's upper surface is a row of small, black glands. The young twigs and buds also are hairy. The flowers are white and occur in many small groups from the ends of branches. The fruit is a small, purplish-black pome.

FRAGARIA VIRGINIANA
Wild Strawberry
Roadsides, fields | May–June | 6 in.

This plant spreads by runners and has basal leaves on long
stalks which project above the flowers. The leaves are divided
into three, sharply toothed leaflets. The white flowers are clustered on separate stalks which also have one or a few small,
uncut leaves. The flower has numerous pistils set into small
depressions on a dome-shaped hypanthium. The fleshy hypanthium is the edible portion of the fruit while the pistils ripen
into pitlike achenes. In moist areas, such as along streams, the
flower and leaf stalks are taller than usual.

PRUNUS SEROTINA
Black Cherry
Roadsides, edge of woods / May–June / 80 ft.

The black cherry is a tree with many (twenty or more) small, white flowers clustered in a raceme. The cuplike center of the flower is smooth and the triangular sepals have few or no glands along their margins. The simple leaves are saw-toothed, with blunt teeth, and the fruit is purple or black. *Prunus virginiana* (choke-cherry) is similar but is generally a small shrub. The marginal teeth on its leaves are longer and more finely tipped, and its fruits are red. *Prunus pensylvanica* (pin cherry) is a shrub or small tree with white flowers grouped together in umbels of two to five flowers. Its leaves are saw-toothed with rounded teeth and there is a gland at the base of each tooth and petiole. The sepals are smooth, but the outer surface of the petals is hairy near their base.

ROSA CAROLINA
Carolina Rose
Roadsides, fields, common along the
seacoast / June–August / 3 ft.

This rose is a shrub with thorny stems, pinnately compound leaves with coarse teeth, and prominent stipules. The flowers have five long and tapering sepals, five broad and unfused pink to red petals, many stamens, and many small hairy pistils. The hypanthium is hollow and ball-shaped and surrounds the ovaries. The sepals, petals, and stamens are attached to the rim of the hypanthium, and glandular hairs are present on the sepals, hypanthium, and on the flower stalks. There are several other common roses, all having pink flowers. *Rosa virginiana* (Virginian rose) is a large shrub (6 ft.) whose leaves have coarse teeth, but it differs in the shape and position of its thorns: in *R. carolina* the thorns are straight and slender and there are many thorns on the stem area between the nodes; in *R. virginiana* the thorns are thick and curve downward and the stem area between the nodes has no thorns. *Rosa palustris* (swamp rose) (6 ft.) and *Rosa nitida* (bristly rose) (3 ft.) have leaflets with fine teeth. *R. nitida* has numerous internodal thorns, while *R. palustris* has few or no thorns between the nodes. *Rosa blanda* (thornless rose) is either thornless or has only a few thorns on the stem.

BLACKBERRIES AND RASPBERRIES

The genus *Rubus* is such a large and highly variable group that even professional botanists cannot easily define the different species. "Brambles" is one term commonly applied to this group, since most of them are spiny shrubs. New canes arise from a permanent underground system of stems and roots and they live for about two years. The first-year canes are generally not branched and do not flower. In their second year, the canes send out short, side branches which flower and fruit. The fruit is a druplet, many of which are grouped together on a moundlike receptacle. Raspberry druplets can be easily separated from the receptacle while blackberry druplets are attached strongly to the receptacle. Another point of distinction between these two subgroups of *Rubus* is that raspberries have round stems and blackberries have angular stems.

RUBUS ALLEGHENIENSIS
Common Blackberry
Fields, open woods / May–July / 9 ft.

This is one of our most common blackberries. The stems have
strong spines, and the young stems and undersurface of the
leaves are covered with glandular hairs. The compound leaves
have three to seven leaflets and the white flowers occur in
racemes.

RUBUS ODORATUS
Purple Flowering Raspberry
Along borders of woods / June–August / 6 ft.

This raspberry is a many-branched shrub. It lacks thorns but has glandular hairs on the new growth of the stems, on the stalks of the leaves and flowers, and on the sepals. The bark of the older stems tend to peel. The leaves are simple, broad, and have three to five lobes. The purple flowers occur in many small groups and the fruit is red and dry.

RUBIACEAE   *Madder family*

This large family of mostly tropical plants includes economically important plants such as those which provide coffee and quinine, and decorative plants such as the gardenia. Our representatives of this family are herbs and shrubs whose opposite or whorled simple leaves have smooth margins. The flowers are regular and their parts are usually in sets of four. The petals are fused and the fruit is a capsule, drupe, or berry.

CEPHALANTHUS OCCIDENTALIS
Buttonbush
Swamps, along rivers / July–August / 6 ft.

Buttonbush is a shrub with small white flowers crowded on very distinctive ball-shaped heads. The flowers are tubular with four fused petals and the style of each flower projects about the length of the flower beyond it. The styles with their stubby stigmas resemble small matchsticks radiating from a round ball. The four stamens lie within the corolla tube. The sepals form a four-lobed, short tube. The broad, uncut leaves are opposite or occur in small whorls of three. The dry fruits split into small nutlets.

461

Galium is the genus to which the bedstraws belong, and there
are several common species of *Galium* which grow abundantly
among or over other vegetation in open fields and woods.
They are slender plants with square stems, whorls of leaves
along the stem, and tiny white or greenish flowers. The small
fruit has two round segments, each with one seed. The fruits
may be bristly or smooth surfaced. Species with bristly fruit
include the following: *Galium circaezans* (wild licorice) is an
erect plant with much broader leaves than other bedstraws; the
hairy, oval leaves occur in whorls of four, and the few greenish
flowers are scattered along long stalks [shown]. *Galium aparine*
(spring-cleavers) grows over other plants and has small spines
on its leaves that cause it to cling easily to clothing; its leaves
are narrow, variable in size, and arranged in whorls of eight;
the flowers are white and occur in groups of two or three at
the ends of small branches. *Galium triflorum* (sweet-scented
bedstraw) also is spiny and grows flat over other plants, but
its leaves are broader and arranged in whorls of six. Species
with smooth fruits include the following: *Galium mollugo*
(great white bedstraw) has erect stems, narrow leaves in whorls
of six to eight, and white flowers in long, branched inflorescences. *Galium asprellum* (rough bedstraw) is a prickly, many-
branched, decumbent plant with its main leaves in whorls of
six and the leaves of secondary branches in whorls of four to
five; white flowers are grouped in branch inflorescences along
the upper parts of the plant.

HOUSTONIA CAERULEA
Bluets
Fields / April–June / 8 in.

Each light blue to white flower of bluets has a yellow center
and is borne singly on a slender stem. The plants generally
crowd together and often cover large areas of a field. There is
a basal rosette of leaves and a few smaller, paired leaves on the
stem. Two types of flowers are produced: one type has a long
pistil which projects beyond the anthers; another type has a
short pistil which lies below the anthers in the corolla tube.
The inside of the corolla tube may be hairy.

MITCHELLA REPENS
Partridge Berry
Woods | June–July | Ground Cover

The partridge berry is a favorite terrarium plant because of its colorful dark green, patterned leaves and red berries. It is a small, creeping plant that can form mats over the forest floor. The leaves are round and evergreen with white veins, and they occur in pairs on the stem. The flowers are white, often pinkish, and are also borne in pairs from the tips of branches. The flowers are funnel-shaped, densely hairy within, and fragrant. The fruit results from the fusion and growth of the ovaries of the paired flowers. Like the bluets, there are two types of flowers: one with a long style and another with a short style.

SALICACEAE *Willow family*

The willow family of trees and shrubs has separate male and female plants. The flowers appear in catkins, in early spring before the leaves appear. The willows have erect catkins while those of the poplars hang downward. Seeds and bracts of the catkins often have long, soft hairs. The fruit is a capsule.

About twenty Massachusetts species of willows have been reported, nine of which occur commonly. The distinctions among the different willows are based mainly on leaf characteristics. Some of the common willows are as follows:

I. Leaves without marginal teeth—the leaves of *Salix humilis* (small pussy-willow) are broader above the middle and have wavy margins; the undersides of the leaves have gray hairs.
II. Leaves with marginal teeth
    A. Undersurface of leaves is hairy and gray—the leaves of *Salix bebbiana* (long-beaked willow) are wide and have coarse teeth.
    B. Undersurface of leaves is hairless and green—the leaves of *Salix cordata* (heart-leafed willow) are not leathery; the stipules are prominent and heart-shaped; twigs are covered with gray hairs [shown]. The leaves of *Salix fragilis* (crack-willow) are long and narrow like those of *S. nigra*, but the leaf bases taper while those of black willow are round. The leaves of *Salix lucida* (shining willow) are leathery, taper to a very long point, and the leaf stalk bases have prominent glands. The leaves of *Salix nigra* (black willow) are long and slender, with fine teeth and round bases. *Salix rigida* (wand willow) is similar to *S. cordata* but has hairless leaves and twigs.
    C. Undersurface of leaves is white—*Salix discolor* (large pussy willow) is a prominent harbinger of spring with its furry

catkins; its elliptical leaves are not hairy, and they are toothed only above the middle. Leaves of *Salix sericea* (silky willow) are hairy and narrow with fine teeth along their margins.

POPULUS TREMULOIDES
Quaking Aspen
Open areas, especially cleared land / April / 50 ft.

Poplar trees also are known as aspens or cottonwoods. They are rapidly growing trees with ovoid to triangular leaves. The quaking aspen is a slim tree with a smooth, greenish bark. The leaves are rounded and egg-shaped and have a flat petiole which catches the slightest breeze to make the leaves quiver. The bracts below the individual flowers are dissected in different ways and may be hairy. *Populus alba* (white poplar) has round leaf stalks and twigs with a dense cover of white hairs. Its bark is smooth and white on the upper parts of the trunk and rough and blackish on the lower parts of the trunk. *Populus grandidentata* (bigtooth aspen) has flat leaf stalks and leaves with large marginal teeth. The bark is smooth and greenish, the twigs are hairless, and the young leaves have white hairs on their lower surface.

SANTALACEAE   *Sandalwood family*

A tropical family of diverse plants; only one representative is found in Massachusetts.

COMANDRA UMBELLATA
Bastard Toad Flax
Woods, clearings / June–July / 12 in.

The bastard toad flax is a parasite on the roots of different woody plants. Its leaves have rounded tips and smooth margins and its white flowers occur in small groups at the ends of branches. Each flower has five sepals, five stamens, and a cup-shaped hypanthium around the ovary. There are no petals and the fruit is a nut.

SARRACENIACEAE   *Pitcher-plant family*

This is an unusual family of insect-catching plants with basal
clusters of hollow, vase-shaped leaves. Only one representative
of this family grows in Massachusetts.

SARRACENIA PURPUREA
Pitcher Plant
Bogs / June–July / 8 in.

Small insects become trapped in the pitcherlike leaves of this
plant. The insects drown and as they decay, their chemical
elements, in particular nitrogen, are used by the plant. The
leaves generally grow in a circular cluster and are partially embedded in the mossy carpet of the bog. The top of each leaf
has a large, erect, fanlike flap with prominent, red veins and
stiff, downward-pointing hairs. The inner wall of the pitcher
is glassy smooth—that, plus the hairs, makes it difficult for a
trapped insect to climb out. A long flower stalk emerges from
the middle of a circle of pitchers and can reach over a foot
high. The nodding flower has five sepals, five reddish-purple
petals which fall off soon after the flower opens, and numerous stamens. The single pistil has a large ovary and a broad,
umbrella-shaped style. The style has five lobes and from each
lobe projects a small stigma.

SAXIFRAGACEAE   *Saxifrage family*

This family of herbs and shrubs has regular flowers. Each flower has five sepals, five petals, ten stamens, and a pistil made up of two separate carpels. The fruit is a follicle. The family contains such familiar shrubs as mock orange, hydrangea, currants, and gooseberries.

SAXIFRAGA VIRGINIENSIS
Early Saxifrage
Woods, rocky ledges / April / 12 in.

The white flowers of early saxifrage are clustered at the ends of small branches at the top of the flower stalk. Usually, a single, hairy flower stalk emerges from a basal cluster of hairy leaves. There are a few reduced leaves on the flower-bearing branches. *Saxifraga pensylvanica* (swamp saxifrage) blooms in May and grows in swampy areas. It has basal leaves and a long —up to thirty inches—hairy, and mostly leafless flower stalk. The flowers are small, with narrow, yellow-green or red-purple petals, and orange anthers.

SCROPHULARIACEAE   *Figwort family*

The figwort family of herbs has irregular, two-lipped flowers with five sepals and five petals. The petals are fused into a corolla tube: two petals form the upper lip and three petals form the lower lip of the tube. There are two to five stamens, some not fully formed, and one pistil. The fruit is a capsule. Several members of this family, such as snapdragon and foxglove, are commonly planted as ornamentals.

CASTILLEJA COCCINEA
Indian Paint Brush
Fields, swamps / May–June / 2 ft.

Most of the color of this plant comes from large, three-lobed bracts associated with the flowers in a spike. The bracts are red while the smaller flowers are greenish-yellow. There are four stamens and the leaves generally have three to five narrow lobes.

CHELONE GLABRA
Turtlehead
Wet areas / July-September / 3 ft.

Each flower of turtlehead has five separate, overlapping sepals and three outer bracts which resemble the sepals. The corolla has a white, hoodlike, upper lip and a pinkish, lower lip. The lower lip has an inner, raised, hairy section. There are five stamens: one is small and sterile and four are tall and fertile with flat filaments and hairy anthers. The stamens occur in two pairs according to their height, and the style is long and arches over the anthers. The flowers are large and clustered in crowded spikes at the ends of branches. The leaves are opposite, toothed, and lance-shaped.

GERARDIA PURPUREA
Purple Gerardia
Fields, swamps, along the seacoast
August–September / 3 ft.

This plant has pink flowers, up to one inch wide, which bloom for only one day. The flowers appear to be regular, but the lower side of the flower extends out a bit and reveals the two-lipped nature of the flowers. There are four stamens and the stems are slender, branched, and bear narrow, opposite leaves. Some plants have only a single stem. The sepals are fused into a bell-shaped structure which has five large triangular teeth.

GERARDIA TENUIFOLIA
Slender-leaved Gerardia
Woods / August–September / 2 ft.

The pink flowers of this gerardia are about one-half inch long and are borne singly on long, threadlike stalks on the upper parts of the plant. The flowers last for only one day and have four stamens. The leaves are opposite and narrow and the stems are branched, especially along the upper parts of the plant. The slender branches tend to spread widely.

GRATIOLA AUREA
Hedge Hyssop
Shores of ponds | June–September | 12 in.

The hedge hyssop is a small, creeping plant. Its terminal part is erect with either simple or branched stems. The leaves are lance-shaped, opposite, and lack stalks. The flowers are bright yellow and have only two functional stamens. Masses of this plant may form a lovely yellow border to small lakes and ponds. [shown]

LIMOSELLA SUBULATA
Mudwort
Tidal mud flats | July–September | 2 in.

This small, inconspicuous plant has clusters of narrow leaves and creeping stems. The small, white flowers are borne singly at the ends of stems mixed among the leaves.

LINARIA VULGARIS
Butter-and-Eggs
Roadsides, fields / June–September / 3 ft.

The leaves of butter-and-eggs are grayish-green, narrow, and they are crowded together on the stem. Branches emerge along the upper part of the stem, and at the end of each branch is a cluster of flowers. The flowers are vertical with a thin, downward-pointing spur. The broader of the two lips has a raised, hairy, orange section which extends into the flower as two strips. There are four stamens, two long and two short, under the upper lip. *Linaria canadensis* (old-field toadflax) grows in sandy, waste areas and blooms from May to September. Its blue flowers have no raised section on the lower lip.

LINDERNIA DUBIA
False Pimpernel
Muddy or sandy shores of ponds / July–September / 12 in.

The false pimpernel is a small, branched plant with oval, paired leaves and lavender flowers. Each flower lies at the end of a stalk which arises from a leaf axil. The upper lip of the flower has two small lobes and it is narrower than the three-lobed, lower lip. There are four stamens, two of which have no anthers. The two complete stamens are located in the center of the corolla tube, above the two sterile stamens.

MELAMPYRUM LINEARE
Cow-wheat
Woods / July–August / 18 in.

The stems of cow-wheat are slim and branched with opposite, lance-shaped leaves. The upper leaves have several sharply pointed lobes or teeth at their base, while the lower leaves have no teeth. The four sepals are fused into a tube at their base but remain as four distinct, sharp lobes on their upper parts. The white- and yellow-tinged flowers are borne singly from the axils of the upper leaves. The long and narrow corolla tube widens to form the two lips typical of the family. The flaplike lower lip fits against the hooded, upper lip. There are four stamens whose anthers appear fused; two of the stamens lie against the lower lip and arch into the upper lip where they meet two other stamens. A long style lies under the upper lip.

MIMULUS RINGENS
Monkey Flower
Swamps, along streams / July–August / 3 ft.

The two-lobed, upper lip of the somewhat flattened, blue flower of *Mimulus* flares upward; the lower lip has three broad lobes and a central, hairy, raised portion with a deep groove, which leads to the inside of the closed flower. There are four stamens, two long and two short, and one pistil with a large ovary. The five sepals are fused except at their tips, which are sharply pointed; these sepals form a five-angled structure. The leaves are opposite, paired, and without stalks, and the base of each leaf abuts the base of its pair around the stem. The individual flowers are arranged on long stalks.

PEDICULARIS CANADENSIS
Lousewort
Woods, meadows / May–June / 16 in.

Lousewort is a hairy plant with fernlike leaves and yellow or reddish flowers grouped together in a head. Each flower is irregular and has a hood-shaped, upper lip, four stamens, and a pistil. The lip ends in two spurs which curve over the lower lip. The sepals are fused and tubular. The leaves appear lacy but they are thick; together with the crowded, irregular flowers they make this plant easy to identify.

PENSTEMON DIGITALIS
Foxglove Beard-tongue
Fields / June–July / 5 ft.

The foxglove beard-tongue is a tall plant whose opposite, lance-shaped leaves lack stalks. The lower portions of the stem are purple. White flowers are grouped at the top of the stem in a loosely branched cluster. The corolla tube is trumpet-shaped with a somewhat two-lipped front; the lower lip is streaked with purple lines. There are four fertile stamens, two long and two short, with black anthers and there is one sterile stamen whose upper end is hairy and bristly. The long style is located between the two petals of the upper lip. The flowers and stalks are covered with glandular hairs.

RHINANTHUS CRISTA-GALLI
Yellow Rattle
Fields, roadsides / May–August / 2 ft.

An unusual feature of this plant is the large, pouchlike structure which is formed by the fused sepals. The pouch is vertical-
ly flat when the flowers are open and becomes round when the
fruits mature. The yellow flowers have a hood-shaped, upper
lip and four stamens. The leaves are opposite, narrow and
toothed, and neither the leaves nor the flowers have prominent stalks.

SCHWALBEA AMERICANA
Chaffseed
Woods, fields / July–September / 2 ft.

The yellow- and purple-tinged flowers of the rare chaffseed are borne singly in the axils of hairy, lance-shaped leaves. The leaves are alternate and stalkless and the stem is hairy and unbranched. The sepals are fused into an irregular lobed tube whose lower side is much longer than its upper side. There are four stamens underneath the upper lip of the corolla tube.

SCROPHULARIA MARILANDICA
Figwort
Woods / June–August / 9 ft.

Figwort is a tall plant with numerous small flowers. The flowers are green on the outside and reddish-brown inside and are arranged along long branches. Two of the three lobes of the lower lip of the corolla tube are located in a lateral position while the third lobe bends abruptly downward. There are five stamens, one of which is not fully developed. The leaves are opposite, heart-shaped and toothed, and the stem is grooved and four-sided.

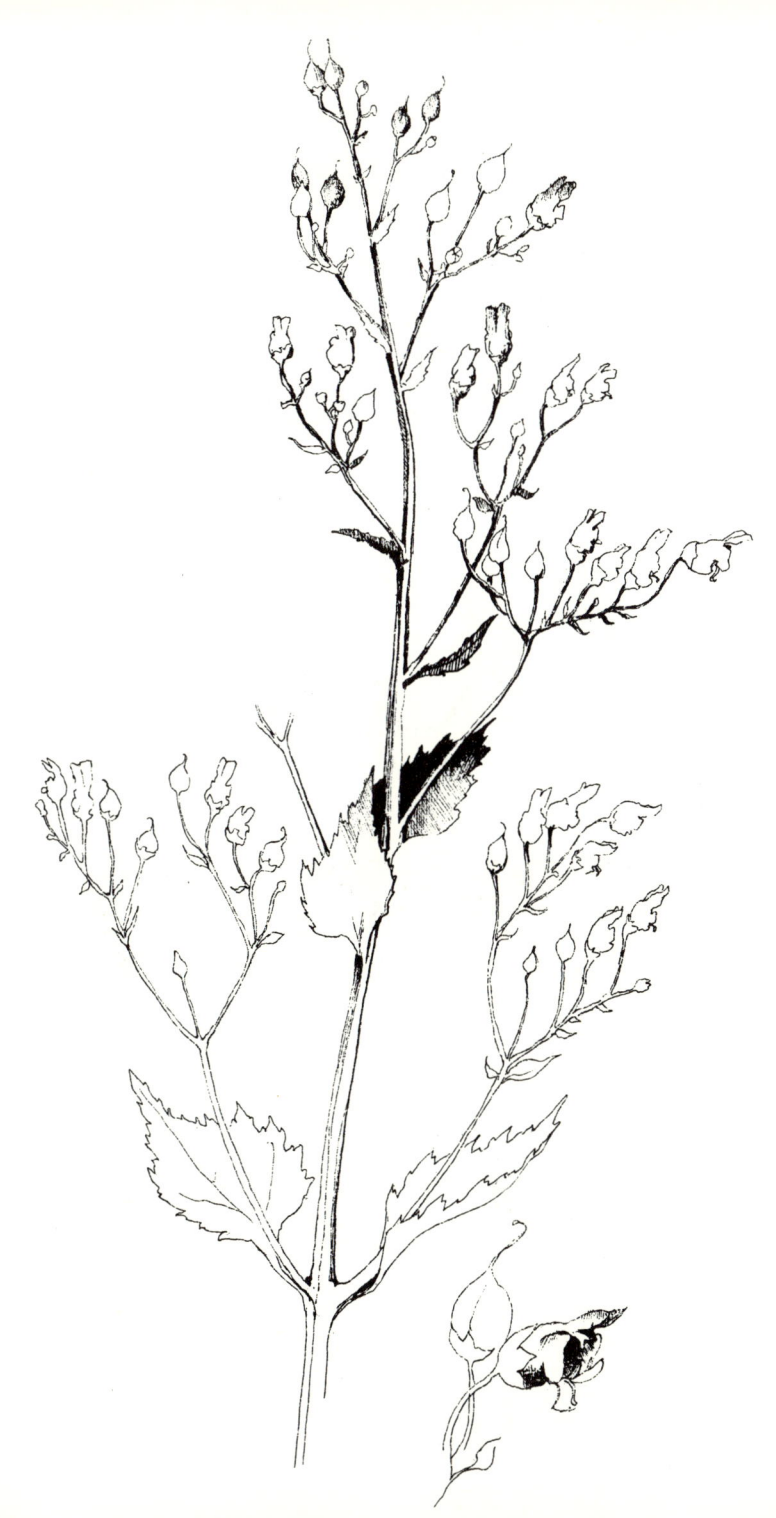

VERBASCUM BLATTARIA
Moth Mullein
Fields, roadsides / July–September / 3 ft.

This biennial plant forms a rosette of leaves in the first year and flowers in the second year. The yellow or white flowers are loosely spaced in a long, terminal raceme. There are five stamens with densely hairy filaments; the positions of the stamens and style are such that the flower bears some resemblance to a moth. The stems and sepals have unbranched, glandular hairs. *Verbascum thapsus* (common mullein) grows up to six feet and has leaves that are covered with white, branched hairs. The yellow flowers are crowded along the top of the stem.

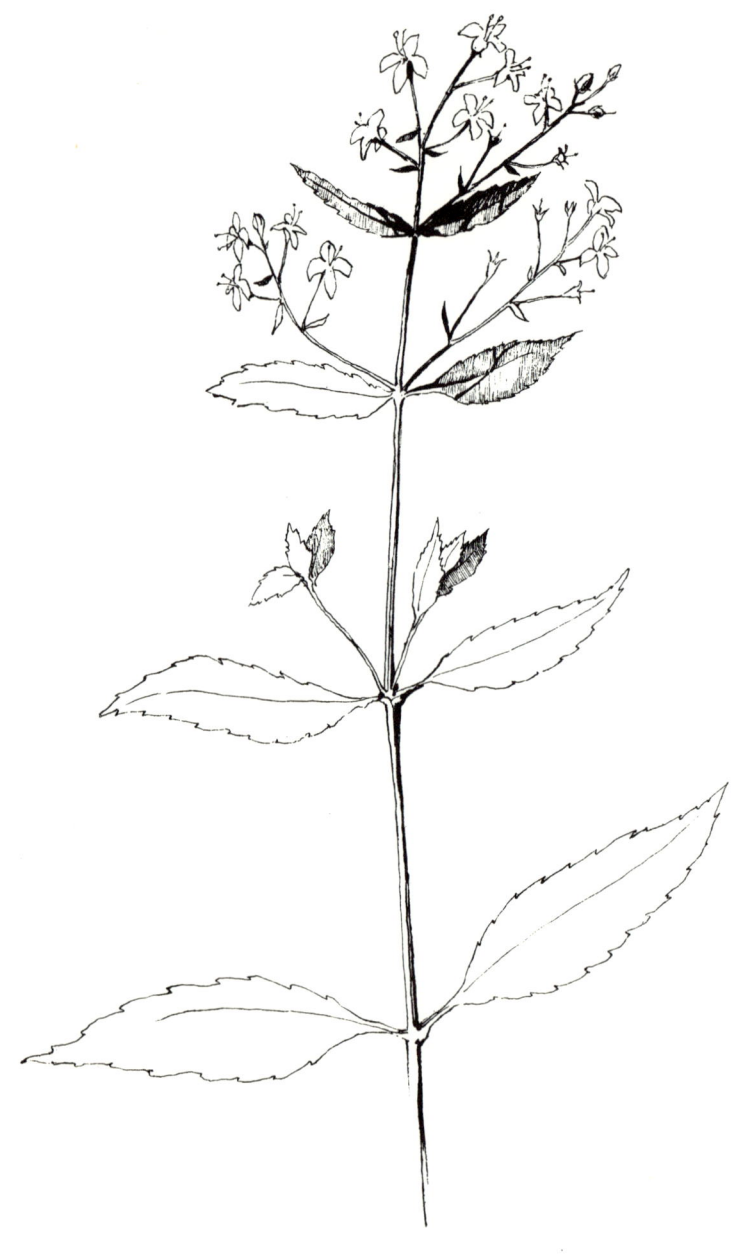

VERONICA AMERICANA
American Brooklime
Swamps, along brooks / June–July / 2 ft.

This creeping, succulent plant has erect tips and bright blue flowers borne in small racemes. The flowers have four sepals and four petals which are fused along their bases; the lower petal is slightly narrower than the other three petals. There are two stamens and one pistil. The leaves are smooth, opposite, and toothed, and are oval- to lance-shaped. *Veronica longifolia* (long-leaved speedwell) is a slender, tall (three feet), erect plant with a long, terminal, crowded raceme of flowers. The leaves have fine teeth and occur in pairs or in whorls of three. The plant grows in gardens, fields, and along roadsides. *Veronica officinalis* (common speedwell) is a hairy, creeping plant whose oval, toothed leaves have stalks; it grows in fields. *Veronica scutellata* (marsh speedwell) is a creeping plant found in swamps and meadows; it has long, narrow, toothless leaves without stalks. *Veronica serpyllifolia* (thyme-leaved speedwell) is a small, creeping plant growing in lawns and along roadsides; it has small, oval, and toothless leaves with stalks.

VERONICASTRUM VIRGINICUM
Speedwell
Woods, meadows / July–August / 6 ft.

Speedwell is a tall, erect plant with sharply toothed, lance-shaped leaves arranged in whorls of three to nine along the stem. The small, white flowers are densely packed in several long spikes at the top of the plant. The two stamens and one style project conspicuously beyond the tubular corolla and make the spikes appear hairy.

SIMAROUBACEAE   *Quassia family*

This family of trees has large, alternate, pinnately compound leaves. Only one species of Simaroubaceae is found in Massachusetts.

AILANTHUS ALTISSIMA
Tree of Heaven
Roadsides, open areas / June–July / 100 ft.

This is a very commonly introduced tree or shrub in waste areas and in cities and towns. It is resistant to urban pollution and most insect diseases and it is a very rapid grower. Its leaves are large and pinnately compound and the leaflets are not toothed. The flowers are small, greenish-yellow and are in loosely branched inflorescences. Bisexual and unisexual flowers may occur on the same tree, the male flowers having an unpleasant smell. The fruit is a twisted samara with one seed in its center.

SOLANACEAE *Potato family*

The potato family of herbs or shrubs has mostly regular flowers. The flowers have five sepals and five petals that are fused
to different degrees. There are five stamens and one pistil and
the fruit is a berry or a capsule. Many representatives of this
family are poisonous, but others yield such well-known compounds as atropine and nicotine and such edible products as
potato, tomato, and eggplant. The common ornamentals *Petu-
nia* and *Nicotiana* also belong to this family.

DATURA STRAMONIUM
Thorn-apple
Fields, waste areas / July–September / 4 ft.

Thorn-apple is a highly poisonous, weedy plant with large,
vaselike, white or purplish flowers. The leaves are large and
toothed and the fruit is an egg-shaped, spiny capsule.

LYCIUM HALIMIFOLIUM
Matrimony Vine
Roadsides / June–September / 9 ft.

Matrimony vine is a shrub with long branches and simple, untoothed, grayish-green leaves. On older branches, the leaves
occur in small clusters and thorns may be present. The short-
lived flowers are light purple and the fruits are red berries. The
flowers have petals that are fused along their lower parts and
spread out at their tips. Each flower is borne singly at the end
of a stalk.

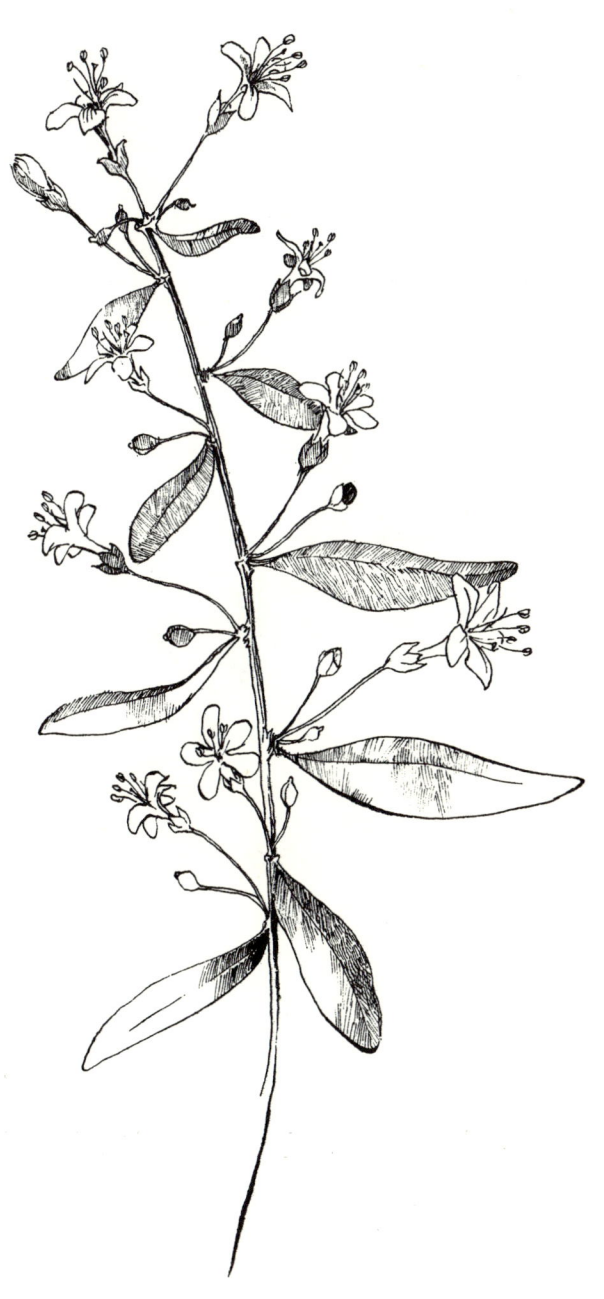

PHYSALIS HETEROPHYLLA
Ground Cherry
Fields, gardens / June–August / 2 ft.

This coarse plant has broad, egg-shaped leaves covered with sticky hairs. The flowers, greenish-yellow with a brown center, hang like bells below the leaves. After flowering, the sepals enlarge greatly and enclose the berry within a lanternlike structure. The berry has many seeds and it resembles a small, green tomato. The plants tend to have a sprawling habit.

SOLANUM DULCAMARA
Climbing Nightshade
Moist, wooded areas / June–August / Vine

The climbing nightshade is a vine with older, woody parts. The leaves are either uncut or have a characteristic pair of earlike lobes at their base. Clusters of purple or white flowers emerge either opposite the leaves or from the internodes. Each flower has five petals that are fused at their base and bent back at their tips. Each petal has two greenish spots near the junction of the free and fused area of the petals. The anthers of the five stamens are fused into a cone around the style; the contrast between the bright yellow cone of anthers and the purple petals is striking. The fruit is a berry, green at first but later turning red. *Solanum carolinense* (horse-nettle) has spines on its stems and leaf stalks; it has white or light-purple flowers, and large, yellow berries. *Solanum nigrum* (black nightshade) has leaves with long petioles, white or light-purple flowers, and black berries.

THYMELACEAE   *Mezereum family*

Only one representative of this mostly African and Australian family occurs commonly in Massachusetts.

DIRCA PALUSTRIS
Leatherwood
Woods / April–May / 10 ft.

Leatherwood is a shrub with simple, untoothed leaves and a very tough, leathery bark which was used for thongs by the Indians. Clusters of small, yellow flowers bloom before the leaves appear. There are no petals, and the sepals are tiny. The hypanthium is long and tubular; eight stamens and a long style project beyond the tube. The fruit is a drupe.

TILIACEAE   *Linden family*

This largely tropical family of trees and shrubs has only the genus *Tilia* represented in Massachusetts.

TILIA AMERICANA
Linden, Basswood
Woods / July / 80 ft.

The linden is a large tree with heart-shaped leaves having unequal sides at their base and clusters of fragrant, yellow flowers. The flower stalk is attached to a leafy, strap-shaped bract which detaches from the tree and helps to disperse the small, nutlike fruits after they mature. The flower has five sepals, five petals, and many stamens.

ULMACEAE   *Elm family*

This family of trees has simple, saw-toothed leaves and clusters of reduced flowers which bloom before the leaves appear.

ULMUS AMERICANA
American Elm
Woods, roadsides / April–May / 100 ft.

The American elm is a large tree that is gradually being destroyed by Dutch elm disease. The flowers are small, perfect, yellowish or purplish, and the fruit is a samara with a notched end and a marginal fringe of hairs. There are no petals and the sepals are fused into a bell-shape. The leaves are unequal at their base and the twigs and buds are smooth. *Ulmus rubra* (slippery elm) is a smaller tree (60 ft.) with hairy twigs and buds; there is no fringe of hairs around its samara.

UMBELLIFERAE   *Parsley family*

A distinctive umbel inflorescence readily identifies members of this family. The umbel is usually compound; that is, the first set of branches (umbel) gives rise to a series of secondary branches (umbellets) which bear the small flowers. There are five unfused petals, and a piece of each petal curls inward as a spinelike appendage. The sepals are tiny or missing. The two styles are swollen at their bases to form a stylopodium, and five stamens protrude from each flower. Below each umbel and umbellet there generally is a cluster of bracts (involucre). The fruit consists of two small, one-seeded carpels, called mericarps; these separate when they mature but remain attached at their tops. Each carpel has five or more longitudinal ribs. The stems are hollow and the leaves are usually compound with their petioles forming a sheath around the stem. This is a large family of herbs with economically important representatives such as carrot, parsley, parsnip, dill, celery, and caraway and also very poisonous species such as water hemlock.

CICUTA MACULATA
Water Hemlock, Cowbane
Wet fields, swamps / July–August / 7 ft.

Water hemlock is a tall plant with widespread branches and a thick, smooth, often purplish stem. The leaves are subdivided several times and the linear or lance-shaped leaflets are generally sharply toothed and are separate from each other. There are many umbels at the top of the plant. All parts of the plant are extremely poisonous.

DAUCUS CAROTA
Queen-Anne's Lace
Fields, roadsides / July–August / 3 ft.

The umbellets of this familiar plant occur on long stalks and they consist of small, white or pink flowers. The flowers on the outer margins of the umbellets have a few larger petals than the flowers within the umbellet. The bracts beneath the umbels are usually subdivided into three to five prongs. The leaves are finely divided and fernlike, and the stems are ridged and have small bristles. The fruits also are covered with small bristles. After flowering, the outer umbellets bend inward against each other and over the inner umbellets to form a nest-like structure. The roots are long and carrotlike; this plant is the wild form of our cultivated carrot. Within the center of the inflorescence there generally is a single, dark-purple flower.

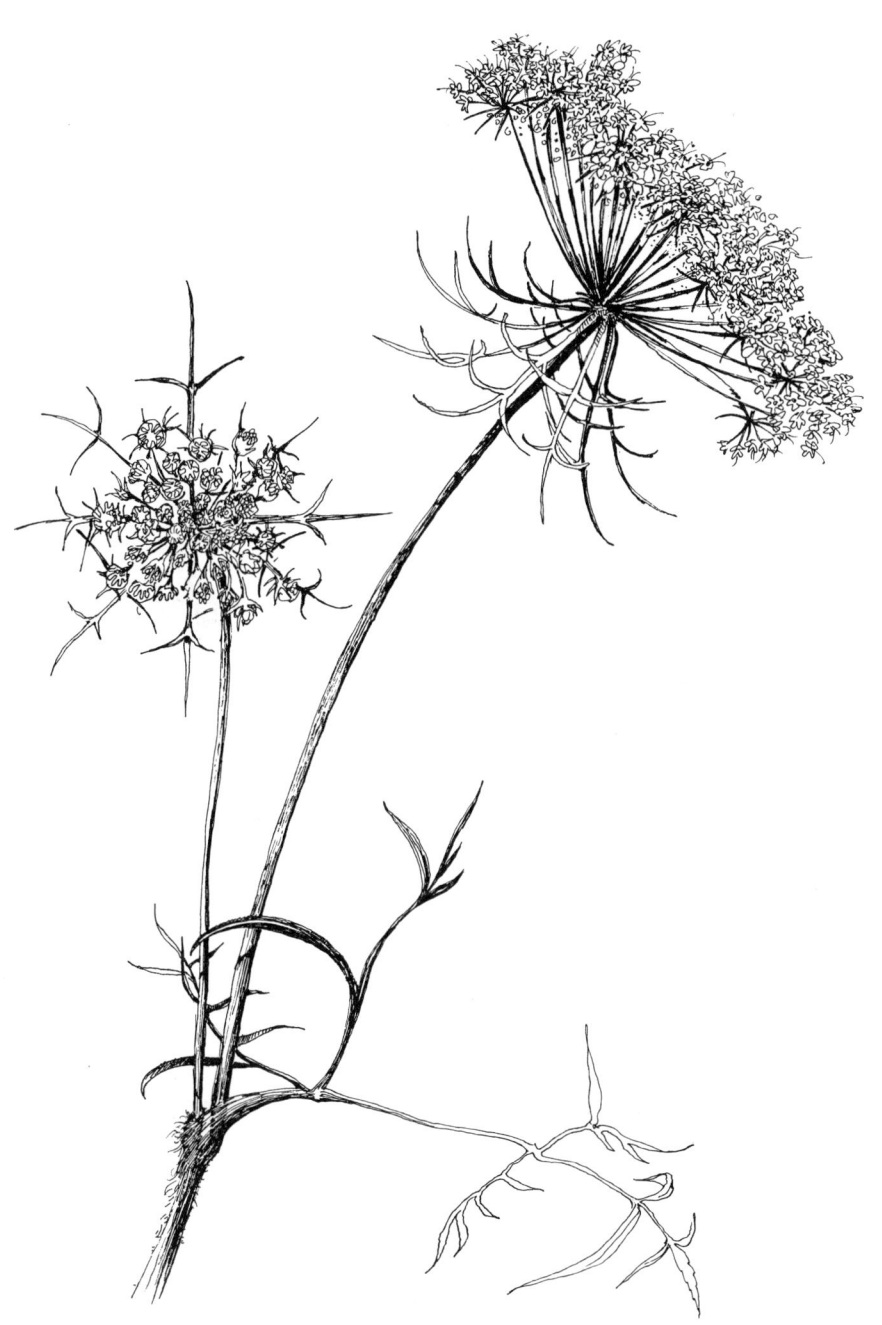

HYDROCOTYLE AMERICANA
Water Pennywort
Wet woods, swamps / July–August / 6 in.

This creeping plant has long, slender stems and round, wavy-margined leaves. The long leaf stalks are attached to the center of the leaf and clusters of minute, white flowers occur in the leaf axils. *Hydrocotyle umbellata* also has circular leaves, but its flowers occur on long stalks and in many flowered umbels. The fruits of these plants are small, round, and flat and they have several ridges along their sides.

SANICULA MARILANDICA
Black Snakeroot
Woods / May–July / 5 ft.

The leaves of black snakeroot are compound with five, sharply toothed leaflets. The small, greenish flowers are in uneven umbels; that is, the primary branches of the compound umbel differ in length. The flowers are either perfect or staminate and the fruit is densely covered with hooked spines. Leaves near the base of the plant have long stalks, while those near the top are much smaller and have short stalks or none at all.

ZIZIA AUREA
Golden Alexander
Meadows, along streams / May–June / 3 ft.

The leaves of golden alexander are divided into three parts, and each leaflet is again subdivided into separate leaflets. The leaf margins are sharply toothed. The flowers are small and yellow and the stem frequently is reddish. The central flower of each umbellet generally has a shorter stalk than the other flowers in the umbellet.

URTICACEAE   *Nettle family*

The nettle family of herbs has prominently toothed leaves and tiny, green, unisexual flowers which lack petals. The fruit is an achene.

BOEHMERIA CYLINDRICA
False Nettle
Swamps, along brooks and ponds / July–September / 3 ft.

The leaves of false nettle are opposite and egg-shaped and have long petioles and coarse teeth. The small, greenish flowers are found in many separate, ball-like clusters on long spikes. The spikes may have small leaves at their tips; one spike may contain either staminate or pistillate flowers, or may have a mixture of both types of flowers. There are no stinging hairs on this plant.

HUMULUS LUPULUS
Hops
Wooded areas, along streams / August–September / 30 ft.

Hops is a perennial vine with opposite, deeply lobed leaves.
There are separate male and female plants, with the male
flowers in a multi-branched inflorescence and the female
flowers in small, conelike clusters. After flowering, the bracts
associated with the flowers enlarge and enclose the fruits. The
dried inflorescences are used to flavor beer.

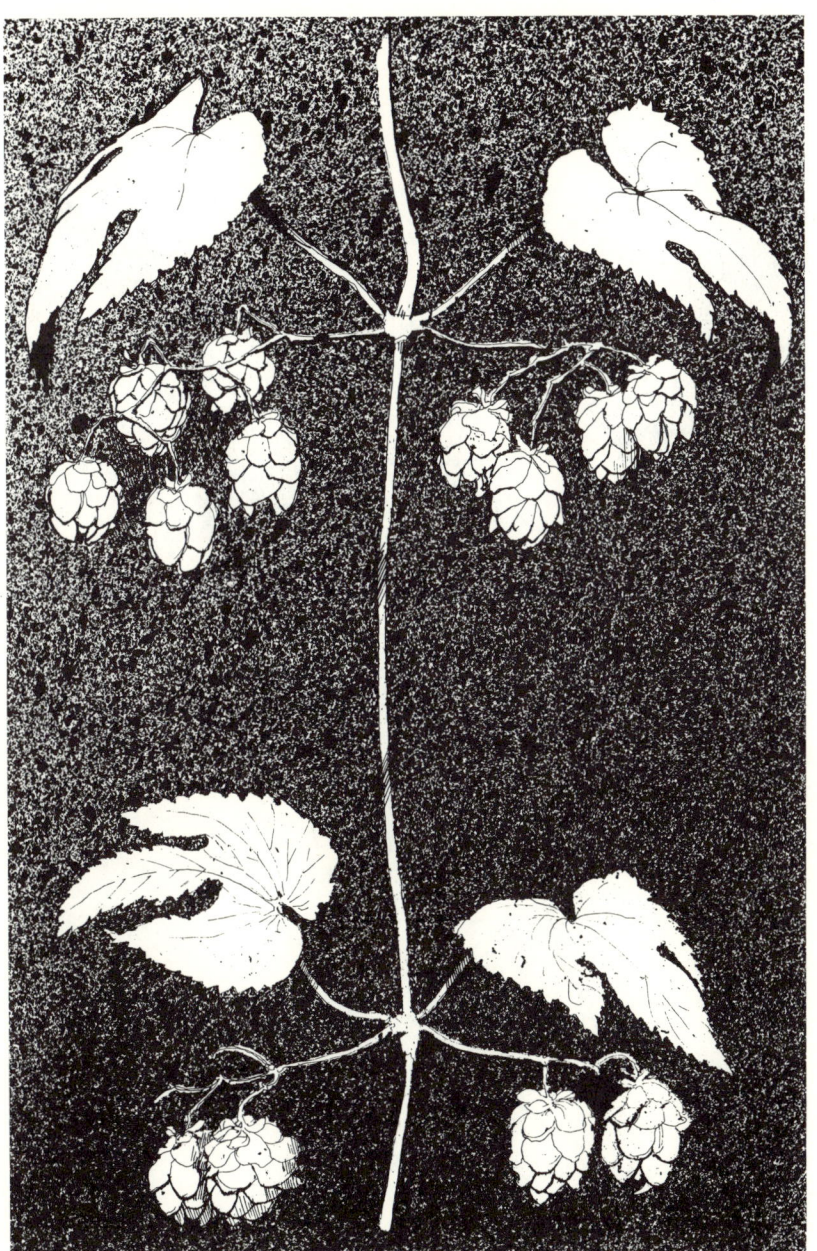

LAPORTEA CANADENSIS
Wood Nettle
Woods / July–September / 3 ft.

The stems and leaves of wood nettle bear stinging hairs. The alternate, egg-shaped leaves are on long stalks. Male and female flowers occur in separate groups on the same plant; the female flowers are multi-branched inflorescences along the top of the plant while the male flowers are in inflorescences nearer the base of the plant.

PILEA PUMILA
Clearwood
Wet, shady areas / August–September / 18 in.

The underside of the egg-shaped leaves of clearweed is shiny and whitish. The stem is clear and smooth and sometimes branched. The flowers are in numerous clusters; separate male and female stalks emerge from the axils of the leaves along most of the plant. Large populations of this plant often grow in shady areas of gardens and yards.

URTICA GRACILIS
Nettle
Roadsides, waste areas / July–October / 6 ft.

The stems and leaves of this nettle are covered with stinging hairs. The leaves are opposite and lance-shaped and the many flowers are grouped on stalks which emerge from the leaf axils. The male and female flowers are on separate stalks on the same plant. If this plant cannot be avoided, it should be handled with extreme caution, because contact with the hairs can be extremely painful.

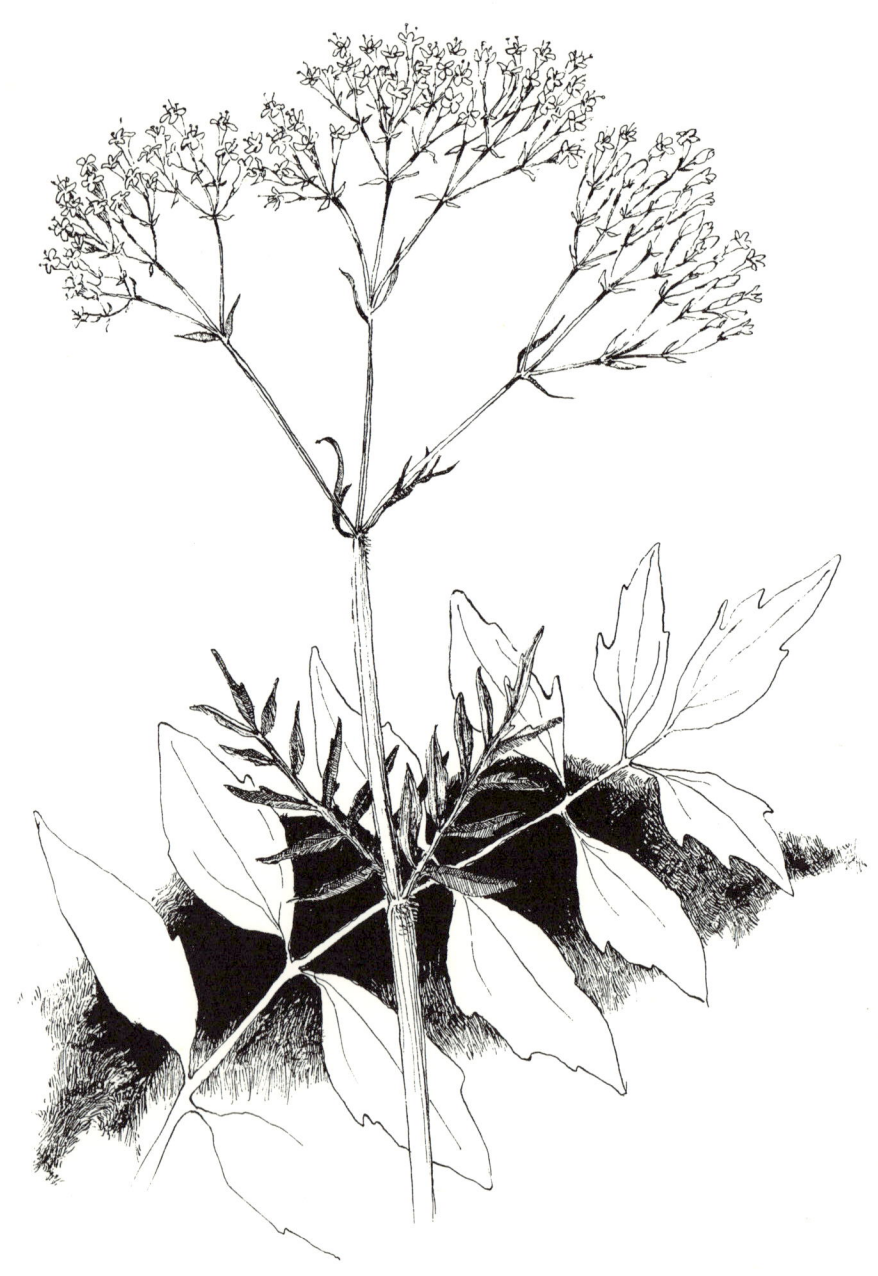

VALERIANACEAE   *Valerian family*

The valerian family of herbs has opposite leaves and small, crowded flowers. The sepals are tiny, or represented by hairs, and the petals are fused into a tube or funnellike shape with five lobes. There are three stamens. The small, dry fruit has one seed.

VALERIANA OFFICINALIS
Garden Heliotrope
Roadsides / June–July / 5 ft.

This common garden plant has escaped from cultivation and has established itself as a wildflower. The plant is hairy with pinnately compound leaves, a thick ridged stem, and fragrant, pinkish flowers in widely branched inflorescences at the top of the plant.

549

VERBENACEAE   *Vervain family*

This large family of diverse plants is mostly tropical and subtropical in distribution. The few representatives in Massachusetts are herbs with opposite leaves, a tubular calyx and corolla, and four stamens (two long and two short). The dry fruit splits into several nutlets when mature.

VERBENA HASTATA
Blue Vervain
Wet areas, along shores and wet fields / July–August / 5 ft.

The blue vervain is a tall plant. Its lance-shaped, hairy leaves have prominent teeth. The basal leaves sometimes have two lateral lobes. The stem is square and furrowed; pairs of smaller leaves occur in the axils of the large leaves. Small blue flowers are in spikes at the end of the stem and each spike has only a few flowers in bloom at the same time. *Verbena urticifolia* (white vervain) has small, white flowers on long, slender spikes.

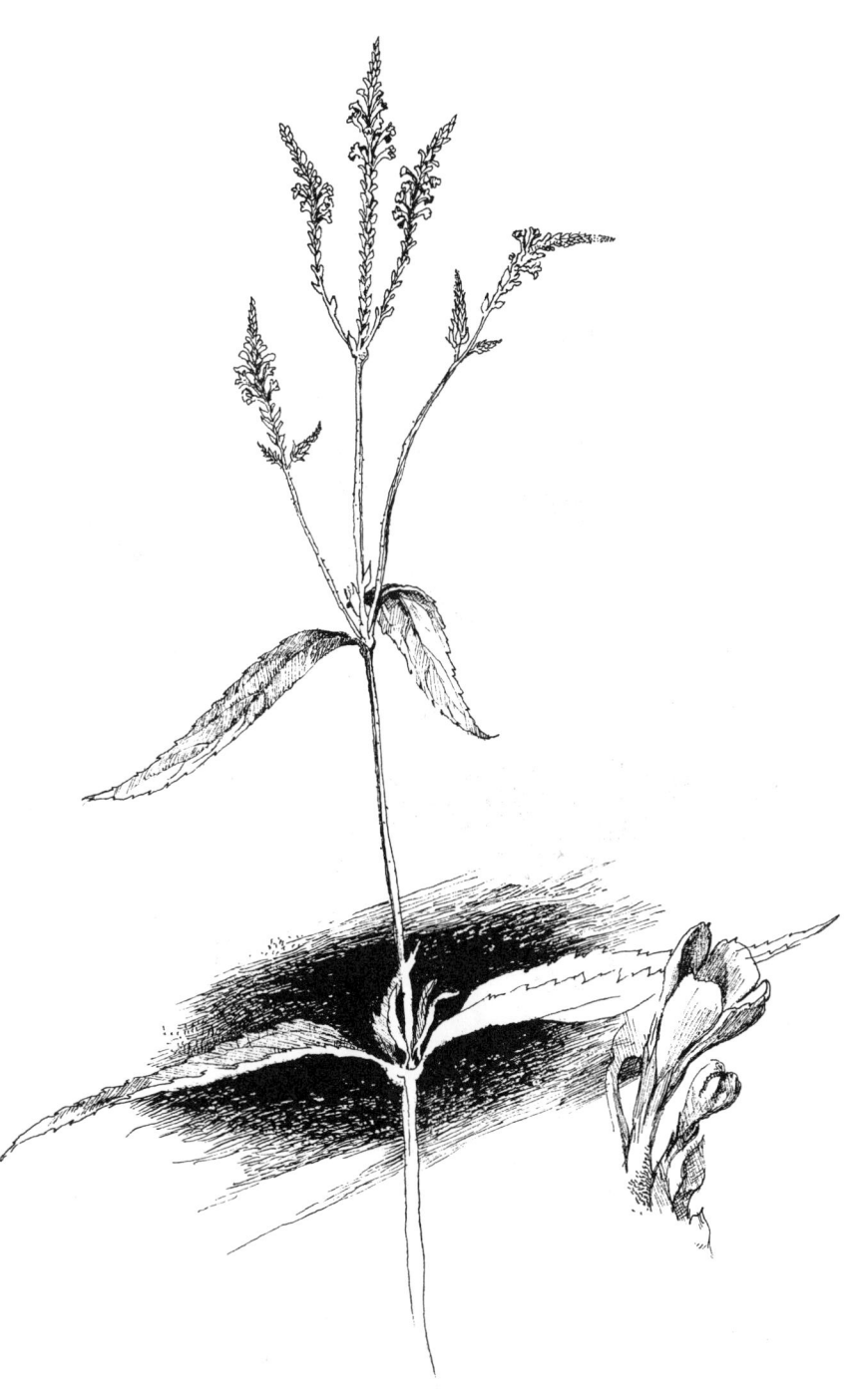

VIOLACEAE *Violet family*

This diverse family of plants thrives in tropical and temperate regions. Our representatives are all herbs that belong to the genus *Viola*. Violets have the following characteristics: Flowers are irregular with five sepals and five petals. There are two lateral petals, two on top, and a lower one which projects backwards and forms a pouch. The lateral petals and sometimes the lower one have patches of hair around the opening of the flower. There are five stamens with broad filaments which extend beyond the anthers. The anthers encircle the ovary, and the style and stigma project beyond the anthers. Each of the two lower stamens has a footlike appendage, and the single pistil has a stigma of variable shape. When the normal flowering period ends, another type of flower, much less visible, appears at the base of the plant near the ground. These flowers do not open, but they self-fertilize and form abundant seeds. The seeds are released after the stalk of the fruit elongates and the capsule breaks open.

Fourteen species of violets occur commonly in Massachusetts, but the distinctions between some of them are trivial. Most of the flowering occurs in May and June. The violets are subdivided into one group that is stemless—that is, the flowers and leaves arise from an underground rhizome—and into another group with stems above the ground.

I. *Stemless Violets.* Two distinctive characteristics of this group are the color of the flower and the shape of the leaves. All the stemless, blue violets, except *Viola pedata* (bird's foot violet), have entire, uncut leaves. The common, blue violets include the following species: *Viola cucullata* (blue marsh violet) grows in swamps and wet woods and has blue, sometimes white flowers whose stalks are longer than the leaves [shown on p. 552]; the flower has a pattern of dark blue veins near its center. *Viola fimbriatula* (ovate-leaved violet) has oval and hairy leaves with round teeth along the margins and indented bases; it is found in dry woods and fields. *Viola papilionacea* (common blue violet) grows in woods and has broad, heart-shaped, smooth leaves; its flowers are different shades of blue but are sometimes white or grayish-white; the flower stalks are shorter than the leaves. *Viola sororia* (woolly blue violet) is similar to *V. papilionacea*, but its leaves and stalks are hairy. *Viola pedata* (bird's foot violet) grows in dry, sandy areas, and is easily recognized by its leaves which are divided into three to five narrow segments; it does not form the small, secondary flowers that other violets produce.

The stemless white violets propagate by means of surface runners or stolons and their flowers have purplish veins. The species include the following: *Viola blanda* (sweet white violet) grows in wet woods and has fragrant flowers, reddish leaf stalks, and shiny, heart-shaped leaves. *Viola lanceolata* (lance-leaved violet) has long, narrow, erect leaves and it grows in swamps, wet fields, and along the shores of ponds and streams.

A common, yellow, stemless violet is *Viola rotundifolia* (stemless yellow violet) which grows in evergreen woods and has egg-shaped leaves whose stalks are covered with fine, white hairs.

II. *Stemmed Violets. Viola canadensis* (Canada violet) pre-
fers limestone woods and may grow up to eighteen inches;
it generally has many stems and heart-shaped leaves crowded
together; its flowers are white and yellow inside and purplish
outside. *Viola conspersa* (American dog violet) grows in woods
and has purple flowers and heart-shaped leaves on short stems
which reach up to six inches in height. *Viola pubescens* (downy
yellow violet) grows in woods and has yellow flowers and
hairy stems and leaves; it does not have basal leaves. *Viola eri-
ocarpa* (smooth yellow violet) has basal leaves and smooth
stems and leaves.

VITACEAE   *Grape family*

This family of thornless, woody vines has small, clustered flowers. The fruit is a berry.

PARTHENOCISSUS VITACEA
Virginia Creeper
Wooded areas, along rivers | June–August | Vine

The leaves of this vine are palmately compound with long stalks and five, saw-toothed leaflets. Numerous flowers are produced in clusters at the ends of branches along the length of the vine. Each flower has five, canoe-shaped petals, five stamens, and a short, squat ovary with a very short style. The fruit is small and black and the tendrils have few branches and no sticky discs. *Parthenocissus quinquefolia* is similar but has branched tendrils and sticky discs which enable them to fasten onto objects.

VITIS LABRUSCA
Fox Grape
Woods / June / Vine

The fox grape has broad, leathery leaves whose undersurface
has a reddish, woolly coating. The twigs also are shaggy and
the berries are purple to black. This species is the ancestor of
several cultivated strains of grape including the Concord grape.
Vitis riparia (riverbank grape) grows along the shores of rivers
and has leaves which are generally hairless on their undersurface.

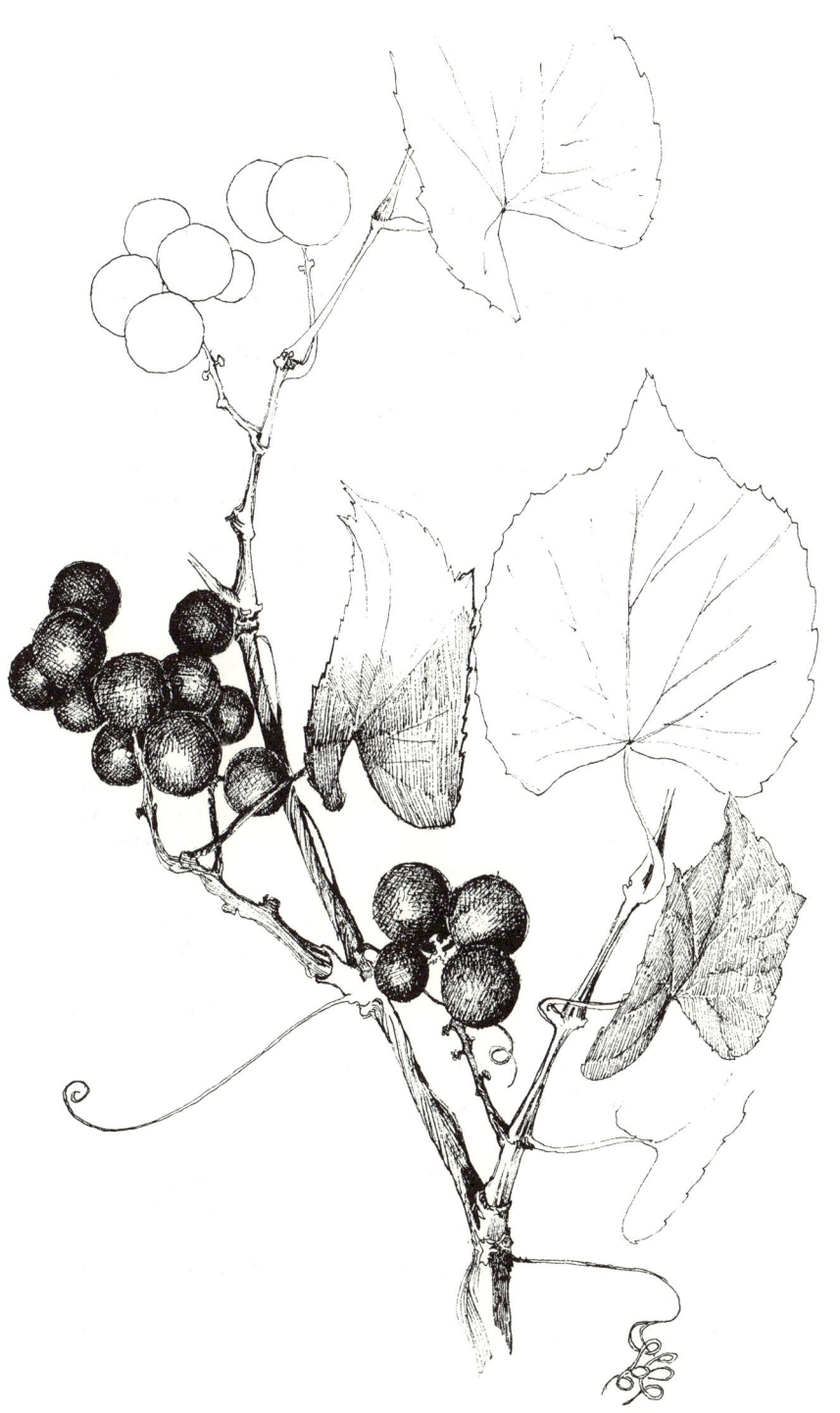

Additional References

Dwelley, M., 1973. *Spring wildflowers of New England.* Down East Enterprise, Inc., Camden, Maine.

Fernald, M. L., 1950. *Gray's manual of botany.* 8th ed. American Book Co., N.Y.

Gleason, H. A., 1962. *Plants of the vicinity of New York.* Hafner Publishing Co., N. Y.

———. 1968. *The New Britton and Brown illustrated flora of the northeastern United States and adjacent Canada.* 3 vol. Hafner Publishing Co., N. Y.

Hines, H. R., and Hathaway, W. A., 1968. *Wildflowers of Cape Cod.* Chatham Press Inc., Chatham, Mass.

Klimas, J. E., 1975. *A pocket guide to the common wild flowers of Massachusetts.* Walker and Co., N. Y.

Klimas, J. E., and Cunningham, J. A., 1974. *Wildflowers of eastern America.* Alfred A. Knopf, N. Y.

Newcomb, L., 1977. *Newcomb's wildflower guide.* Little, Brown and Co., Boston, Mass.

McCann, J; Covey, H. J., Corrinet, R. P.; and Ostroski, P. E., 1972. *An inventory of the ponds, lakes, and reservoirs of Massachusetts.* Community Resource Development Program, Cooperative Extension Service, University of Massachusetts, Amherst (nine reports for the counties of Massachusetts).

Peterson, R. T. and McKenny, M., 1968. *A field guide to wildflowers of northeastern and north-central North America.* Houghton Mifflin Co., Boston, Mass.

Petrides, G. A., 1958. *A field guide to trees and shrubs.* Houghton Mifflin Co., Boston, Mass.

Rickett, H. W., 1966. *Wild flowers of the United States.* Vols. 1 & 2, The Northeastern States. McGraw-Hill Book Co.

Seymour, F. C., 1969. *The flora of New England.* Charles E. Tuttle Co, Rutland, Vt.

Stupka, A., 1965. *Wildflowers in color.* Harper and Row Publishers, N. Y.

Symonds, G. W. D., 1958. *The tree identification book.* Wm. Morrow and Co., N. Y.

———. 1963. *The shrub identification book.* Wm. Morrow and Co., N. Y.

560

Glossary

achene—a small, dry fruit with one seed

anther—one of generally two joined sacs that contain pollen and which together with the filament constitute a stamen

awn—a bristlelike extension of the lemma of the grass floret

axillary flowers—occurring in or originating from the axils of leaves; i.e., the region between the base of the leaf petiole and the stem of the plant

berry—a fleshy fruit with one to many seeds

biennial—a plant that undergoes a two-year cycle, with vegetative growth the first year and flowering and fruiting the second year

bract—a specialized and usually modified leaf located at the base of a flower or inflorescence

bulb—a tuberous, underground structure which is actually a modified stem enveloped by fleshy leaves

calyx—a collective name for the sepals, or outermost set of floral leaves, of a flower

capsule—a dry fruit formed from a compound pistil that opens by pores or by splitting

carpel—a specialized leaf that forms a simple pistil

catkin—an erect or, more usually, a hanging inflorescence of closely packed and reduced unisexual flowers associated with bracts

column—in the orchid flower, a structure formed by the fusion of the stamens and styles

deciduous—characterized by loss of leaves during the fall season

dicotyledons—a large class of flowering plants whose flower parts are in sets of four or five or multiples of these and whose leaves have a network of veins

drupe—a fleshy fruit with one seed contained in a stony pit

druplet—a smaller version of a drupe, usually one of several which are grouped together

floret—one of several small flowers that make up a spikelet in grasses, and one of many flowers that make up the head inflorescence in the composites

follicle—a dry fruit formed from a single pistil and splitting along one side only

561

glume—a specialized bract found at the base of a grass floret

head—the characteristic inflorescence of the daisy family (Compositae); the florets are packed together on the same receptacle

hypanthium—an expansion of the receptacle which may be cup-shaped, saucer-shaped, or tubular-shaped

inflorescence—the specific pattern of flowers on a plant; for example, a spike, umbel, raceme, and head

involucre—a close group of bracts which occur below an inflorescence in some plants

irregular flower—all the sepals and petals are not the same size or shape; the flower has a bilateral symmetry

lemma—in the grasses, one of two bracts which enclose the flower and which bears the awn

lenticels—small, wartlike dots or circles on the stems and branches of trees and shrubs; they correspond to air pores

mericarp—in the parsley family (*Umbelliferae*), a one-seeded fruit; they occur in pairs because the two carpels of these plants split but do not separate completely

monocotyledons—a class of flowering plants whose flower parts are in sets of three or multiples of three and whose leaves have parallel veins

node—that part of a stem to which the leaves are attached

nut—a dry, one-seeded fruit with a hard outer coat; often encased in a husk

nutlet—a smaller version of a nut

palea—in the grasses, one of two bracts which enclose the flower

panicle—a branched inflorescence

pappus—in the daisy family, the hairs or bristles found on the achenes

perfect flower—a flower with both stamens and pistils

perigynium—a membranous sac around the ovary of the sedges

petiole—leaf stalk

pistil—the female organ of a flower; made up of a stigma, style and ovary; formed from one or several fused carpels

pistillate flower—a flower with only pistils and no stamens; female flower

pith—the generally soft tissue directly in the center of a stem

pollinium—in orchids and milkweeds, a mass of adhering pollen grains

pome—a fleshy fruit, formed mostly from an expansion of the hypanthium

raceme—an inflorescence that consists of a central, elongated stem on which are located flowers with stalks; the oldest flowers are on the lower part of the stem

receptacle—the expanded end of the flower stalk on which are borne the floral parts *regular flower*—the sepals and petals are alike in size and shape; the flower has a radial symmetry

rhizome—an underground stem

samara—a dry, one-seeded fruit with a winglike extension or margin

sepal—an outer part of a flower

spadix—in the arum family (Araceae), the thick, clublike inflorescence

spathe—a highly modified or colored bract associated with the spadix in the arum family

spike—an inflorescence that consists of a central elongated stem on which are located flowers without stalks

spikelet—in the grasses, the smallest unit of the inflorescence

stamen—the male organ of a flower; made up of usually two anthers and a stalk (filament)

staminate flower—a flower with only stamens and no pistils; a male flower

stigmatic rays—discernible lines or stripes leading from the stigma to the ovary

stipule—a leaflike structure located at the base of a leaf petiole

stolon—a runner or horizontal stem which grows above the ground and forms roots

style—portion of a pistil located between the stigma and the ovary

stylopodium—in the parsley family (*Umbelliferae*), a structure formed by the swelling of the bases of two styles

taproot—the main root of a plant

tendril—a slender extension of a leaf or a stem which a plant uses to fasten onto objects or other plants

tuber—a short, thick, underground stem

umbel—an inflorescence where all the flower stalks arise from a common point

umbellet—a small umbel which is part of a group that makes up a compound umbel

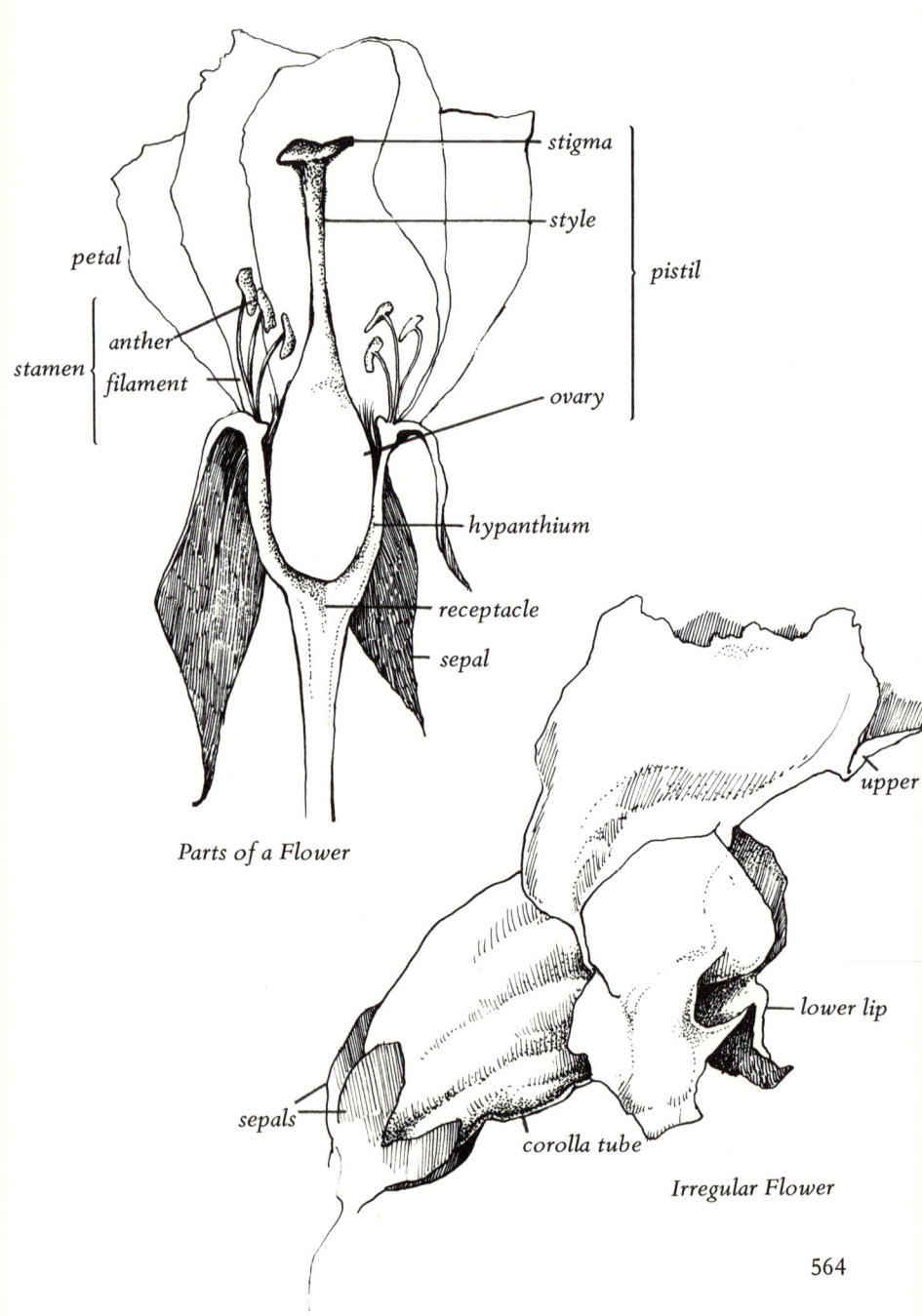

stigma

style

pistil

petal

ovary

stamen

anther

filament

hypanthium

receptacle

sepal

Parts of a Flower

upper

lower lip

sepals

corolla tube

Irregular Flower

564

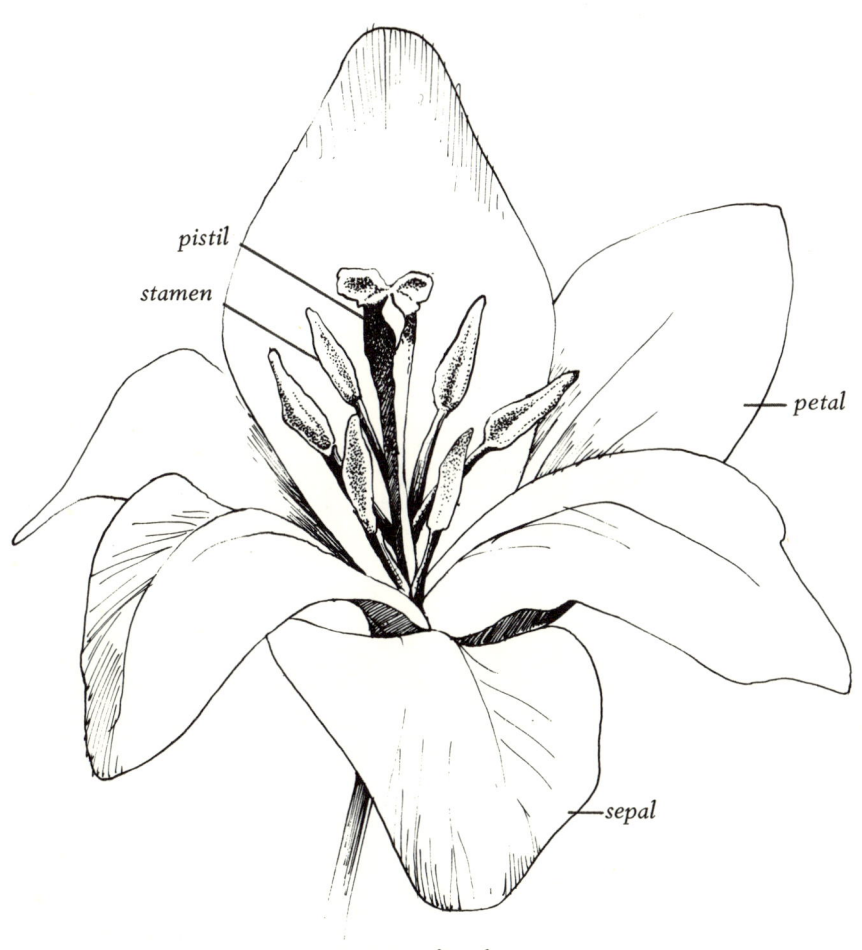

pistil

stamen

petal

sepal

Regular Flower

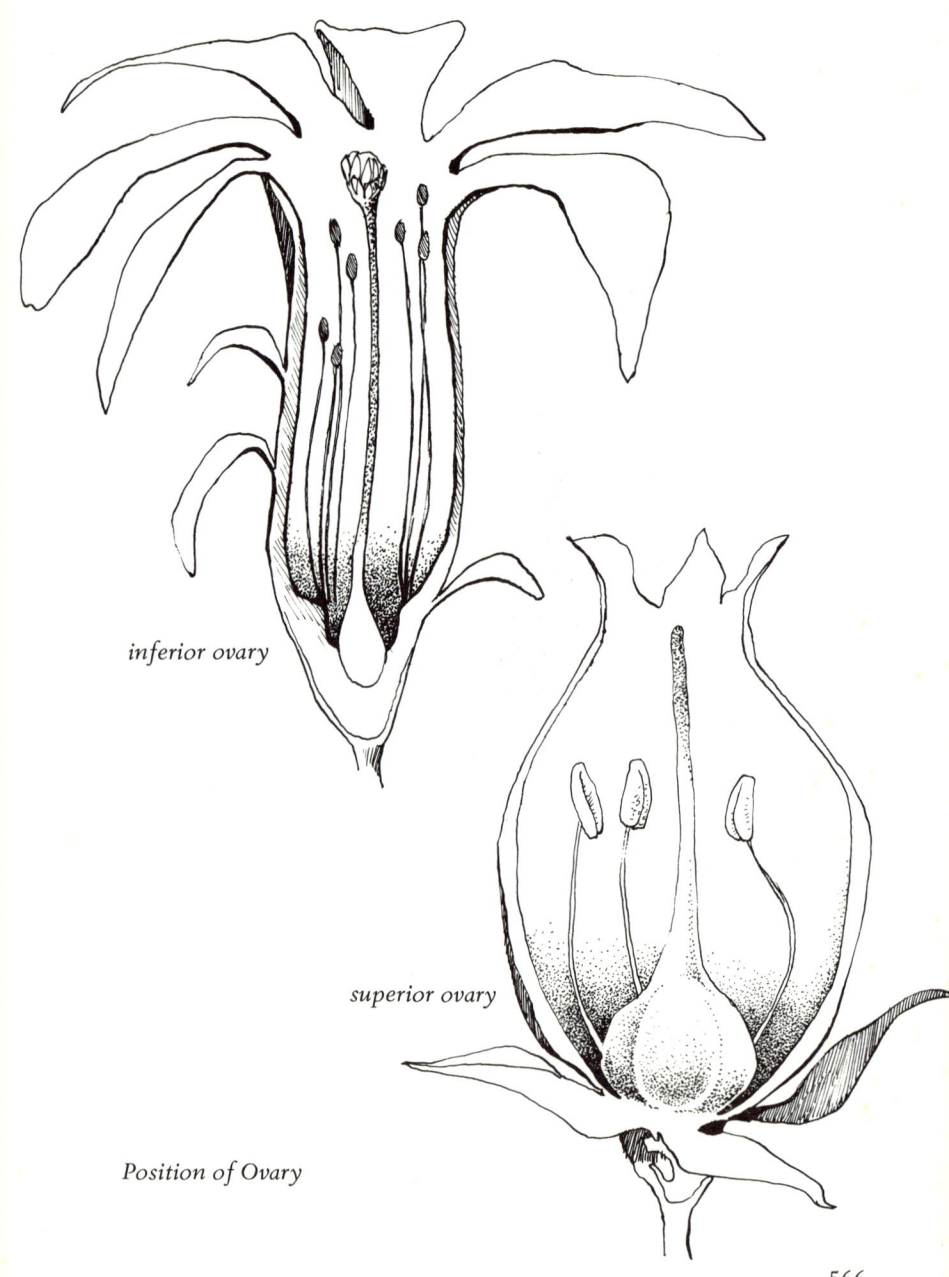

inferior ovary

superior ovary

Position of Ovary

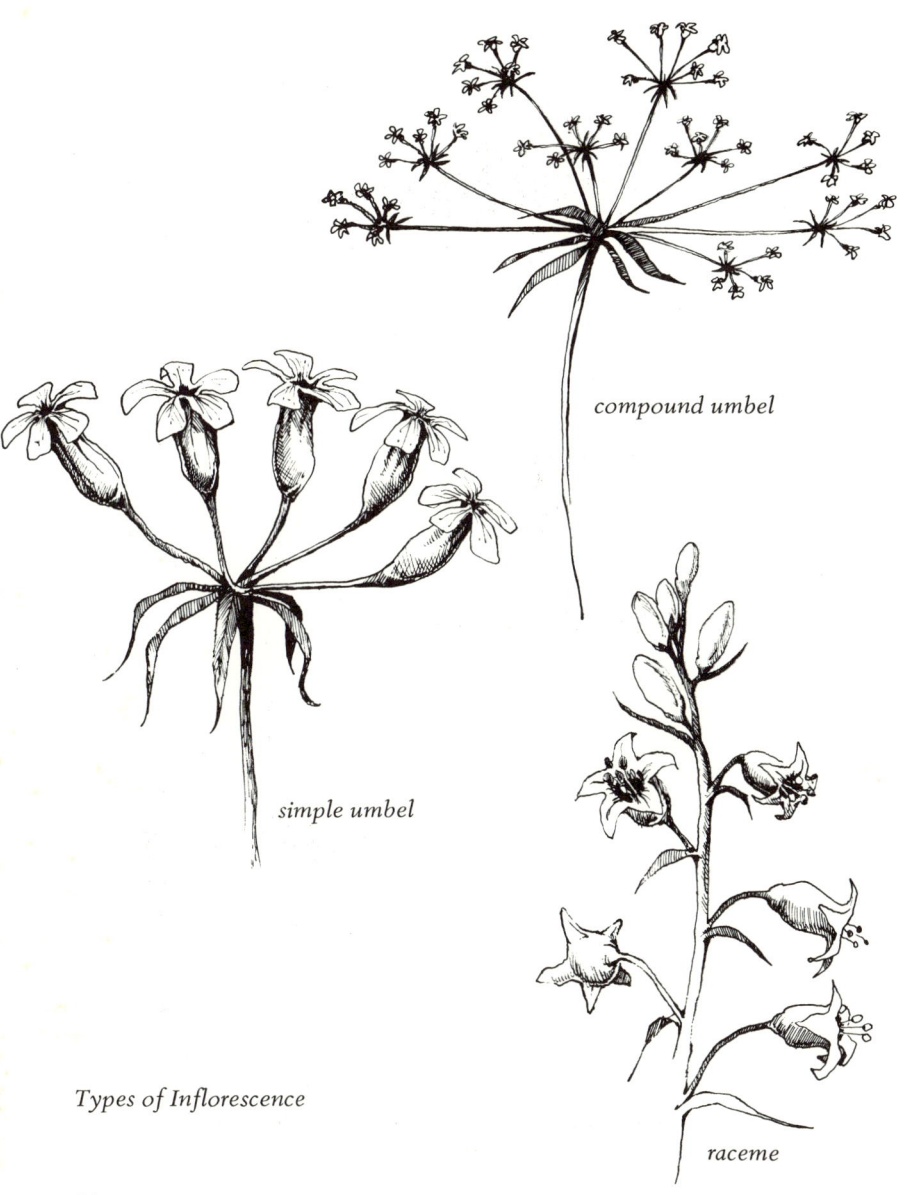

compound umbel

simple umbel

raceme

Types of Inflorescence

567

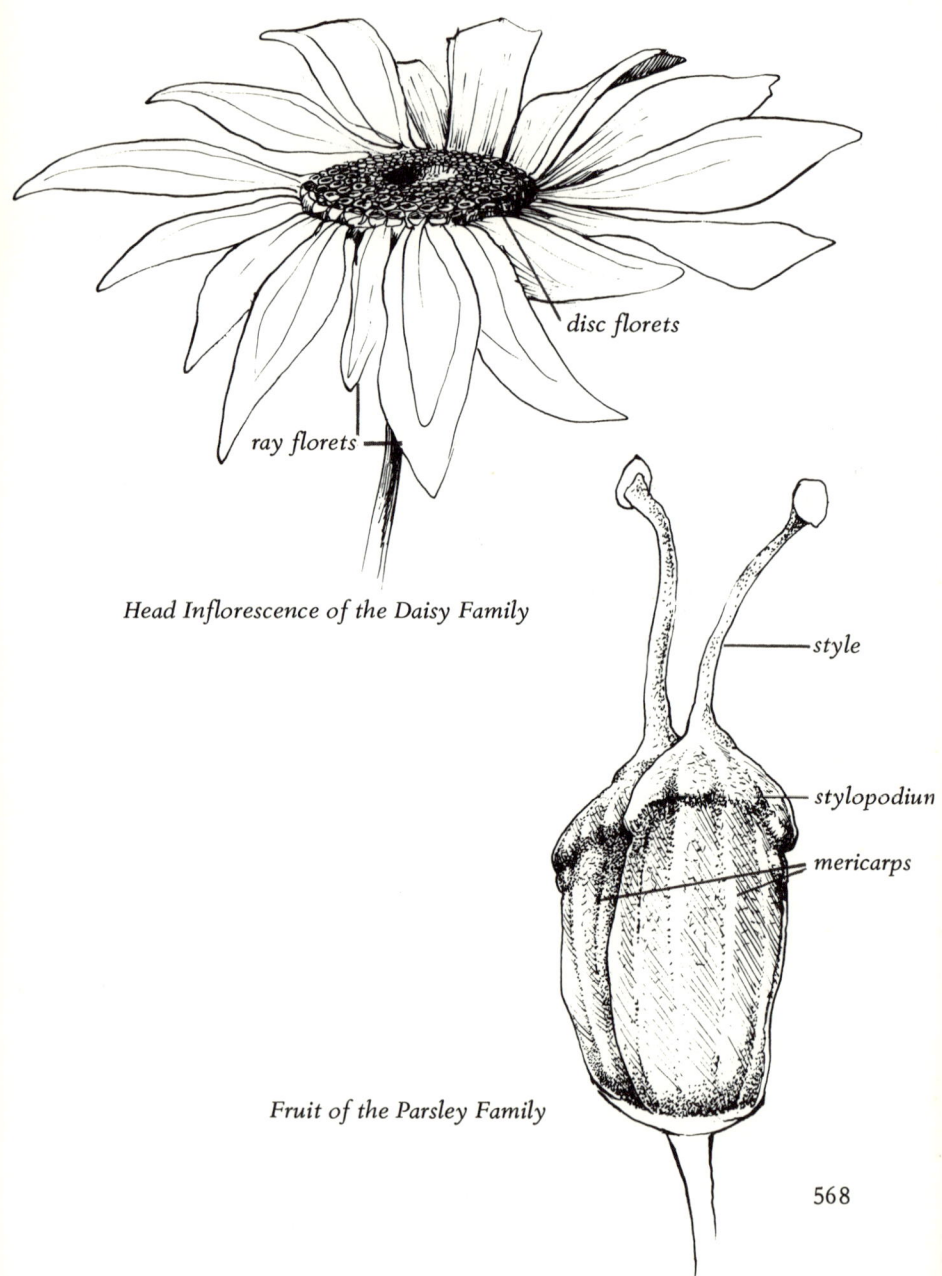

disc florets

ray florets

Head Inflorescence of the Daisy Family

style

stylopodiun

mericarps

Fruit of the Parsley Family

568

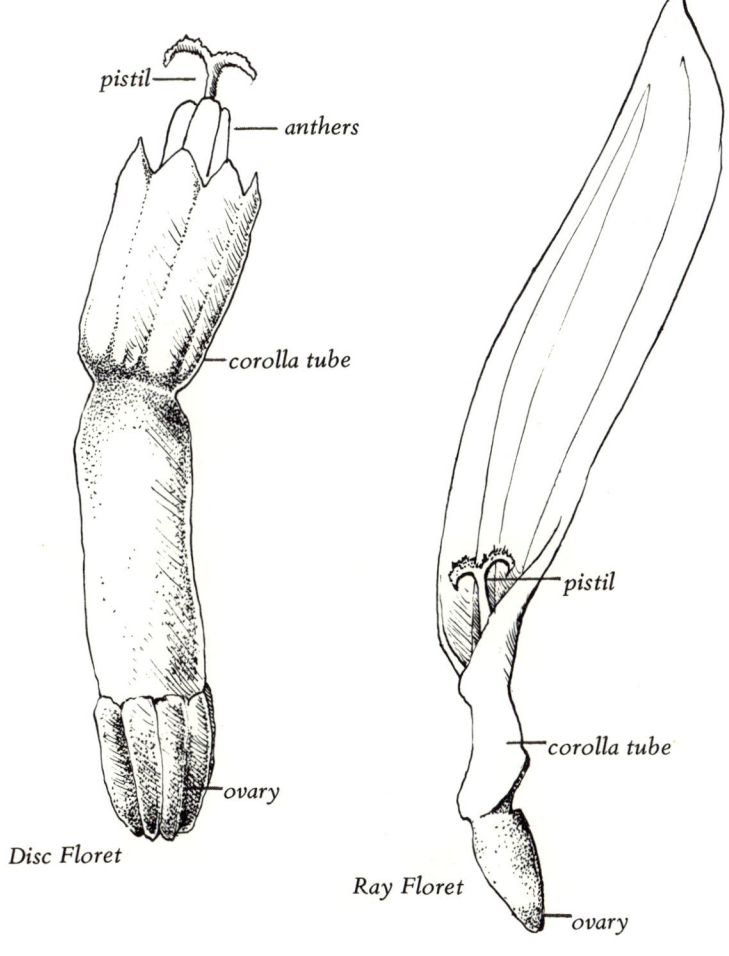

pistil

anthers

corolla tube

ovary

Disc Floret

pistil

corolla tube

ovary

Ray Floret

Parts of a Head Inflorescence

569

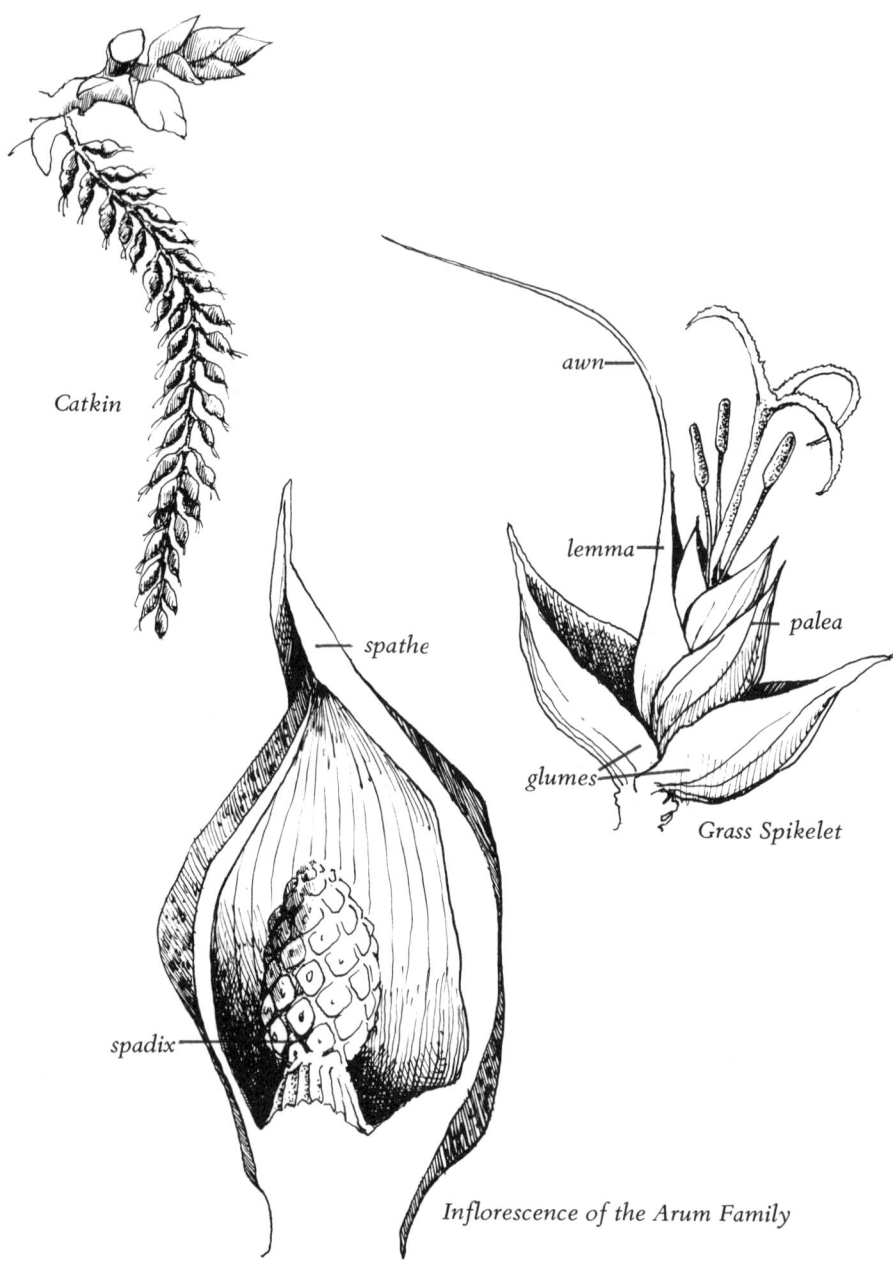

Catkin

awn

lemma

palea

glumes

Grass Spikelet

spathe

spadix

Inflorescence of the Arum Family

Index

Page numbers for illustrations are shown in italics.